M. T. Gerin Lajoie

Dyspepsia and How to Cure It

Description of the Digestive Organs and of the Different Kinds of Food

M. T. Gerin Lajoie

Dyspepsia and How to Cure It
Description of the Digestive Organs and of the Different Kinds of Food

ISBN/EAN: 9783337811747

Printed in Europe, USA, Canada, Australia, Japan

Cover: Foto ©berggeist007 / pixelio.de

More available books at **www.hansebooks.com**

DYSPEPSIA

AND HOW TO CURE IT.

DESCRIPTION OF THE DIGESTIVE ORGANS AND OF THE DIFFERENT KINDS OF FOOD.

SPECIAL CHAPTERS ON OBESITY, LEANNESS AND THE SKIN.

BY

M. T. GERIN-LAJOIE, M. D., C. M.

Member of the "New Hampshire Medical Society," Member of the "Societe Francaise d' Electrotherapie," of Paris; Member of the ' Societe d' Hypnologie" of Paris; etc., etc.

SIXTH EDITION.

NASHUA, N. H.
TELEGRAPH PUBLISHING COMPANY,
1895.

AUTHORS REFERRED TO IN THIS BOOK.

P. Tillaux.—Typographical anatomy Professor at University (Paris).

Gray.—Anatomy.

Thenard.—First to extract "osmazone" from meat.

Chevalier and Baudrimont.—On wines.

Hunter.—About obesity.

Pidoux.—Rapport de l'herpetisme et des dyspepsies (Paris).

Brocq.—Maladies de la Peau (Paris).

Ewald.—Diseases of the stomach (Berlin).

Dieulafoy.— Pathologie Interne (Professor at University), Paris.

Hartshorne.—Essentials of the Principles and Practice of Medicine (Philadelphia).

Dujardin-Beaumetz.—Journal de Therapeutique, 1880, p. 828.

W. Jaworski.—Munichener Med. Worchensch., 1887, No. 33.

Austin Flint.—Flint's Practice of Medicine (Philadelphia) 1881, p. 477.

Rossier.—Merycisme hereditaire, etc. Annal. de la Soc. de Med. d'Anvers, Avril-Mai, 1887.

F. Kretschy.—Deutches Archiv. fur Klin. Med., Bd. 18, S. 257.

E. Fleischer.—Berl. Klin. Worchensch., 1882, No. 7.

Ewald and Boas.—Virchow's Archiv. Bd. 104.

H. Kirsh.—Dyspepsia uterina. Berl. Klin. Worchenschi, 1883, No. 18.

Rossbach.—Gastroxynsis.

Lepine.—"Nervous Gastroxia." Bull. de la Soc. Med. des Hopitaux, 10 Avril, 1885.

Germain See.—Des dyspepsies gastro-intestinales, 1883 (Paris).

W. Fenwick.—Virchow's Archiv. 1889, Bd. 113, S. 189.

Louis.—De la Tuberculose.

Hutchinson.—The Morbid States of the Stomach and Duodenum (London, 1878).

Rosenthal.—Berliner Klin. Worchensch., 1888, No. 45.

Hayem.—Des alterations du Chimisme Stomacal dans la chlorose. Bull. Medical, 1891, No. 87.

Biernaki.—Berl. Klin. Worchensch. 1891, Nos. 25 and 26.

Colleville.—Progress Medical, 1883, No. 20.

Charcot.—Des crises tabetiques, etc., Gazette Med. des Paris 1889, No. 39.

Rosenstein.—Berliner Klin. Worchensch. 1890, No. 13.

Burney Yeo.—On the Treatment of the Gouty Constitution, British Medical Journal, January 7 and 14, 1888.

TABLE OF MATTERS.

PAGE.

Introduction.

PART I.

Life, its chemical definition, 1

CHAPTER I.

§i. Description of the digestive organs:—The mouth and the teeth, their functions. The salivary glands, their use and functions. The saliva. The tongue, and where are the senses of taste, of flavor. etc. §ii. The pharynx, the œsophagus, and their function in digestion. §iii and iv. The stomach, its anatomy, its position and its functions. The muscular (?), the cellulous and the mucous coats and their blood supply, their glands, etc. The nerves of the stomach, right and left pneumogastrics, and the sympathetic nerves. Chymification process. 3

CHAPTER II.

§i. Intestines; division. Small intestines: duodenum, jegunum and ileon. Large intestines: cœcum, colon and rectum. §ii. Small intestine, its form, position and function; glands, chylification. §iii, Large intestine, its position and function. §iv. Digestive juices. §v. Saliva. §vi. Gastric juice. §vii. Pancreatic juice, §viii. Bile. 9

CHAPTER III.

§i. Digestion, its influence over body and mind. §ii. Where food is digested; physiology teaches us that meats and the azotous part of food are digested in the stomach, that fats, oils and fecula are *not* digested or transformed in the stomach, but in the duodenum under the influence of the bile and pancreatic juice, they are divided and absorbed to pass into our blood. Salts in food and where they go. 14

CHAPTER IV.

Classification of food: 1st, Hydrocarbonated and where it goes, its importance to our life; 2nd, Albuminoids, which goes to all our tissues; 3d, Fats and their necessity; A, Who ought to eat food of the first class; B, Who ought to eat food of the second class; C, Where food of the third class goes. Considerations on food of these three classes. Which food lean persons ought to eat; which one fat persons are to avoid. 16

CHAPTER V.

§i. Meats. §ii. Dark red colored meats, their effect upon the system and blood, the different kinds. §iii. Red colored meats, different kinds and their effects. §iv. White colored meats, their action upon the system and blood. 19

CHAPTER VI.

§i. Eggs, composition, kinds, how to eat them, how to find their age. §ii. Milk, its composition, how to take it. §iii. Butter, cheese, fats and oils. §iv. Salt. §v. Bread. §vi. Rice, macaroni, vermicelli. 22

CHAPTER VII.

Vegetables. §i. Potatoes, composition, use. §ii. Sweet potatoes. §iii. Jerusalem artichoke. §iv. Carrots, cabbage, beets, turnips, onions. §v. Peas, beans, corn. §vi. Lentils, squash, tomatoes, olives, cucumbers, celery, salads. 27

CHAPTER VIII.

Considerations on pastry, 32

CHAPTER IX.

Fruits. §i. Pickles, melons (cantaloup), watermelons, dates, figs, dry raisins; prunes, their use in constipation for women especially, peaches, grapes, apples. §ii. Honey is said to insure long life 33

CHAPTER X.

Beverages. §i. Water, its danger. §ii. Different liquors made in each country. §iii. Wines, and those preferred by great men. §iv. 1st class, wines, which act upon the digestion—2nd class, those more alcoholic—3rd class, sweet and generous wines—4th class, champagne, etc. Table of wines and their proportion of alcohol. §v. Alcoholic liquors. §vi. Malt liquors. §vii. Ciders (perry). §viii. Hydromel, how to make it. §ix. Cordials. §x. Consideration on alcoholic liquors and how to fight intemperance. §xi. Vinegar, its effects upon digestion and blood. §xii. Sugar, its action upon the stomach contents. 36

CHAPTER XI.

§i. Coffee, what organ it mostly has effect upon; how to make a good cup of coffee. §ii. Teas, black and green, etc. §iii. Cocoa, where it is found and in what way it is mostly eaten—Chocolate, to whom it is specially adapted. 47

CHAPTER XII.

General considerations about food and constitutions or temperaments. §i. Sanguine temperament: By what constituted. Best food for it. What food to reject. Its danger. §ii. Bilious temperament, By what constituted. Best food for such temperament. Its danger. §iii. Nervous temperament. What constitutes it. What is the best regime. What food to avoid. State of general system when errors of diet are made. 52

CHAPTER XIII.

Obesity—What the Spartans thought of it—What foods are transformed into adipose tissue. What foods to avoid—What foods to eat to cure obesity. What beverages to take, and how to take care of the general system—Baths—Hunter's idea about obesity—Food; Hygienic, and electrical treatments. History of a very interesting case. 55

CHAPTER XIV.

Leanness—Its causes—Food to be taken—What food to avoid—Regime—Baths, and care of the bowels in leanness. 59

CHAPTER XV.

The Skin. §i. 1st, food that has an immediate action—The way it acts—History of an interesting case. 2nd, food of slow action—Hygiene of the skin—What kinds of baths for different skins—How to preserve the freshness of face and shoulders—

How to get rid of blackheads and how to have a nice skin. 61

Table of foods. 68

PART II.

CHAPTER XVI.

Dyspepsia. §i. Persons affected. §ii. Kinds of dyspepsia—Every case of dyspepsia has more or less nervous weakness attending it—What Ewald says about " nervous dyspepsia" or Neurosis of the stomach—How nervous dyspepsia is brought about—What Dieulafoy says about it. §iii. Cause of dyspepsia—What Hartshorne writes about it—Different causes of dyspepsia—Mental strain of every description—Cause pertaining to the teeth, imperfect mastication—Inordinate use of spirits, opium, tobacco, coffee, tea—Dyspepsia with "society people"—Reflex symptom of a diseased organ, such as the brain, spinal column, ovaries, womb, etc—"Dyspepsia uterina," etc. 71

CHAPTER XVII.

Diagnosis. §i. To differentiate between a local or general cause; or if due to a remote diseased organ. §ii, Prognosis—Treatment must be gradual and there are many "rechutes." 79

CHAPTER XVIII.

Symptoms of Dyspepsia—General symptoms—Pain in the stomach—Neuralgia—Vomiting — Merycism Palpitation of the heart—History of two very interesting cases, observations IX and XXI—Craving hunger—"Nervous Gastroxia"—Disorders of the vision—Foul breath—Skin shows symptoms of digestive disturbances—Dyspeptic cough—Constipation—Intestinal dyspepsia. 81

CHAPTER XIX.

Treatment of Dyspepsia—HYGENIC—Temperance in food and drink—Regular habits—Number of meals—Seventeen rules to follow. MEDICAL—Pepsin—Elixir peptogene—Hydrochloric acid—Lavage of the stomach—Bitters—Eleven prescriptions—Danger of Carlsbad and Marienbad waters—Hot milk—Cardialgia—Gastralgia—Its treatment—Pyrosis—Its treatment—In "Nervous diseases" of the stomach—Constant electrical current—Faradic current—Albuminate of Iron—Constipation—Its treatment. ELECTRICAL—Special treatment of the author—Description of treatment—Its effects upon dyspeptics—History of a few cases—Observation VI—Observation XIX—Observation XXXIX. 92

CHAPTER XX.

Gases in the Stomach and Intestines—Gastro-intestinal pneumatosis, or Flatulent Dyspepsia—Gas in the intestines, why?—Food that will bring on gases—Gas is formed by fermentation—by abnormal secretion of the mucous membrane of the stomach and intestines—Symptoms—Belching—Regurgitation—Oppression—Slow digestion—Distention of stomach—Vicious circle—Palpitation of the heart—Diminished mental activity—Sleep and dreams—Treatment. 110

CHAPTER XXI.

Diseases of organs, other than the stomach, causing dyspeptic disturbances. TUBERCULOSIS—Dyspepsia in tuberculosis—ANÆMIA and CHLOROSIS—their dyspeptic symptoms—HEART DISEASE—DISEASES OF THE KIDNEYS—Dyspepsia in floating Kidneys—LIVER—difficulty to find whether the liver or the stomach is first affected. DIABETES—Always examine urine in a middle aged dyspeptic — GOUT — RHEUMATIC DIATHESIS — SKIN—MALARIAL POISONING. 115

Electricity Appendix. 1

INTRODUCTION.

This little book does not hope for perfection; but if these few pages can point out, to those tormented by dyspepsia, the means of relief from its distress, then the author will be rewarded for his work by the knowledge that he has saved his fellow-men from pain.

THE AUTHOR.

ERRATA.

Page 6, line 40. For plate III, read plate IV.
Page 6, line 5. For plate III, page 8, read plate IV, page 10.
Page 8, last line. For plate III, read plate IV.
Page 13, line 12. For plate IV, read plate III.
Page 13, line 18. For plate IV, read plate III.

LIFE.

ITS CHEMICAL DEFINITION.

Life is an unceasing and noiseless combustion—The food is the fuel; the digestive canal represents the laboratory where combustion and the resulting different changes take place. This animal combustion begins with life and continues without interruption until death. Our body is incessantly destroyed and renewed until renewal ceases, then life becomes extinct.

To keep a fire up the burnt combustible must necessarily be replaced, likewise the stomach must be given food to keep up that unknown force which is called Life. Digestion is, therefore the indispensable function, the absolute necessity to keep up the "vital fire."

The word combustible, applied in an indirect way to food, is not a metaphor, as our food contains more or less carbon combined with other principles. Carbon passes in the blood which arrives at the lungs through the pulmonary arteries, there it is burnt by the oxygen of the inhaled air, and comes out at each expiration in the form of carbonic acid. Therefore there is real combustion; and it is by the combustion in the lungs that, always at the same degree, the temperature of our body is kept up in the coldest as well as in the hottest climes.

Without oil the lamp will not burn,
Without food life comes to an end.

CHAPTER I.

ORGANS OF DIGESTION.

To understand well by what mechanism dyspepsia can be brought on, we must know which are the organs that prepare, transform and absorb our food. We will find here the place for a summary description of the digestive organs, their position, functions, etc.

§I. THE MOUTH.

We all know what the mouth is and where it is situated; but very few know how to use the *knives* and *mill-stones* (teeth) which nature has placed there. Teeth are an ornament, but they are especially instrumental in cutting up and grinding the solid food, so that the saliva (furnished by the salivary glands, parotide, sub-maxillary and sub-lingual) will thoroughly penetrate it and prepare it for more important transformation in the stomach.

The sense of taste, of flavor, the action of turning the food ground by the teeth, the action of rejecting foreign substances, etc., are vested in the tongue.

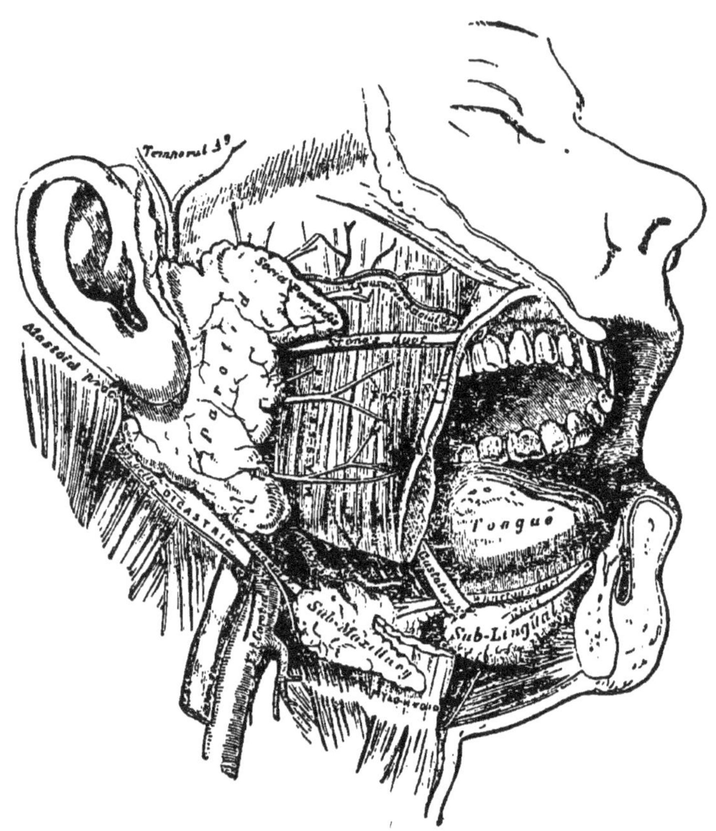

PLATE I. The mouth, teeth, tongue and the salivary glands.

§II. THE PHARYNX AND THE ŒSOPHAGUS.

At the posterior part of the mouth is the pharynx, which receives the food after its passage through the glottis,* the œsophagus's function is

*Glottis is the door at the posterior part of the mouth which closes the larynx, when food is to go to the pharynx.
AUTHOR.

to transmit to the stomach what the pharynx gave up to it. From the lips to the stomach is a tube composed of: The mouth, the pharynx and the œsophagus.

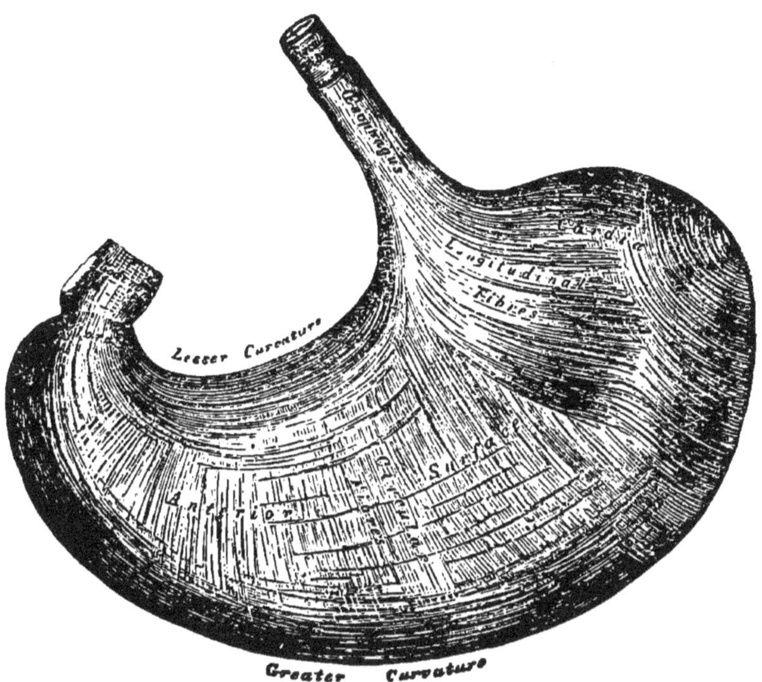

PLATE II. Incorrect position of the stomach.

§III. THE STOMACH.

The most important organ of digestion is the stomach. It is a large sac with two openings; one for the inlet of food and drink, the cardiac orifice; the other for the outlet of the transformed food, the pyloric orifice. Contrary to what is generally believed the stomach does not lie horizontally, as in plate II; but almost perpendicularly, as in plate III.

This position, (plate III, page 8,) of the stomach will explain easily why dyspeptics often have palpitations of the heart. The gâses accumulate in the upper extremity of the stomach and trouble the action of the heart; even sudden death has been caused in this way after a copious meal. In 1892, a young man, 21 years of age, who was troubled with dyspepsia, took quite a solid supper well along in the evening, just before retiring. His companion awoke a few hours after to find his bed-mate dead. An autopsy was made and no heart disease could be found to have caused death, but gases were found to have accumulated in the stomach and brought on a syncope which caused his demise.

The stomach is composed of three different coats:

1st. *The muscular coat* composed of three sets of fibres: the longitudinal, circular and oblique. In these reside the combined contractions which *triturate* the food (chymification process).

2d. *The cellulous coat* connecting the muscular with the mucous coat.

3d. *The mucous coat*, better called the mucous membrane of the stomach. This is the most important of the three; the gastric juice is secreted by glands called the peptic glands which are in this mucous membrane.

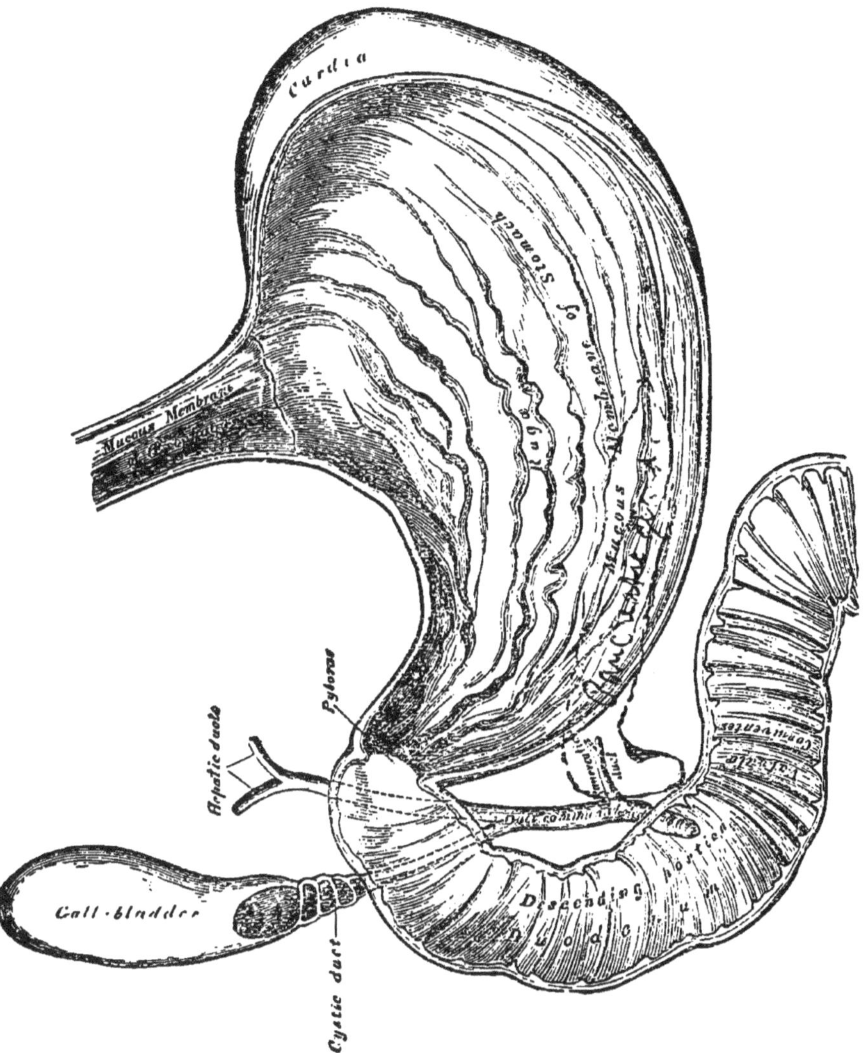

PLATE III. Anterior walls of stomach and intestine taken away, so as to show the interior, and also the relations to the stomach of the pancreatic gland, gall bladder, etc.

Arteries. The blood supply is furnished by arteries: The coronaria, pyloric, gastro-epiploic and vasa brevia arteries. Numerous veins absorb some of the nourishing principles to throw them in the torrent of circulation, for it is in the stomach the digestion of all albuminoids and meat takes place. The food after its transformation in the stomach is called the *chyme*.

Nerves. The nerves presiding over digestion are: The terminal branches of the right and left pneumogastrics, the former being distributed over the posterior and the latter over the anterior part of the stomach; the sympathetic also supplies the organ with a number of branches. These nerves are the principal ones in our body, running down directly from the head, furnishing a few branches to the heart and lungs on their way to the stomach. We can now readily understand why one has a headache when the stomach is over-worked; the digestive disturbance is telegraphed to the head through the nerves, the result being a sick headache.

§IV. POSITION OF THE STOMACH.

The position of the stomach is splendidly illustrated in plate III, and it is self-explanatory.

CHAPTER II.

§I. INTESTINES.—FORM, SITUATION AND FUNCTIONS.

The intestinal tube comprises six different intestines in its length.

Three, the duodenum, the ileon and the jegunum form the small intestine.

The other three, the cœcum, the colon and the rectum, form the large intestine.

For the more easy comprehension of our subject we will leave the subdivision and will speak only of the small intestine and the large intestine.

§II. SMALL INTESTINE.

The small intestine begins at the pyloric orifice of the stomach and ends at the ileo-cœcal valve (which separates it from the large intestine.) It is about 21 feet in length.

Like the stomach it has three coats: The muscular, cellular (or submucous) and the mucous coats; in this last one are many glands and absorbing vessels which aid and complete the process of digestion.

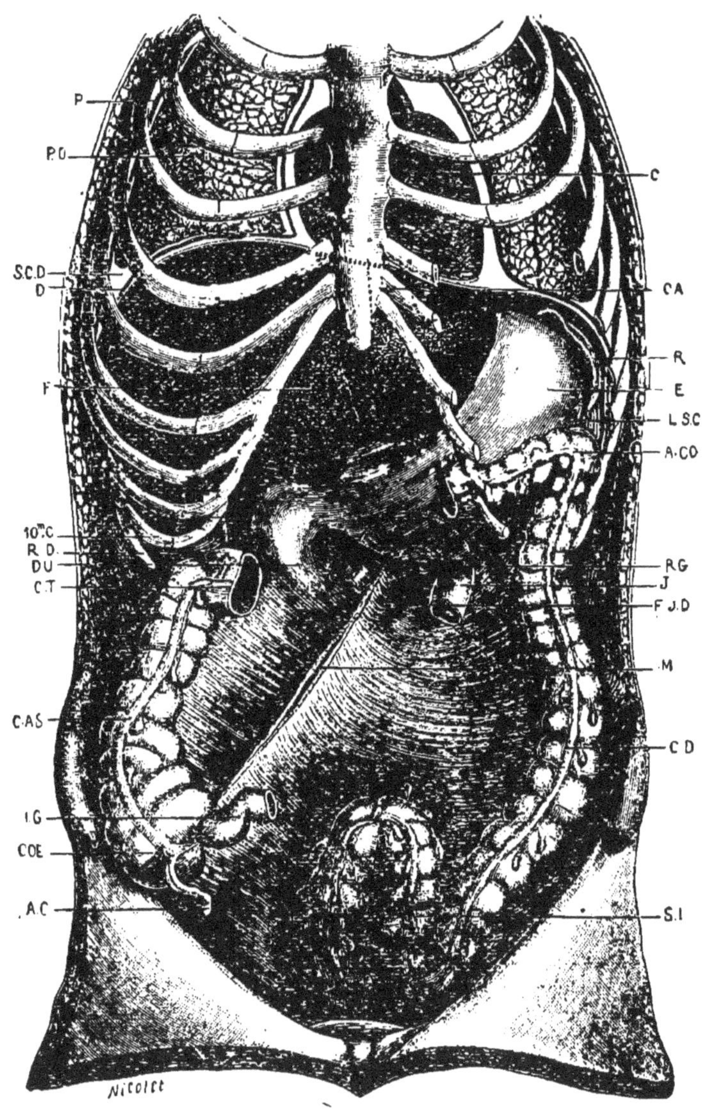

PLATE III. Correct position of the stomach and other organs.

EXPLANATION TO PLATE III.

A. C. Appendix (blind bowel.)
A. C. P. Angle of the colon.
C. Heart.
C. A. Cardia (outlined by dotted lines) of the stomach.
C. A. S. Colon.
C. D. Colon.
C. O. E. Cœcum.
C. T. Colon (part of which has been taken away to show underlying organs.)
D. Diaphragm.
D. U. Duodenum (outlined to jegunum), small intestine.
E. Stomach (the cardiac part of which is inside the dotted lines).
F. Liver, the left lobe covers part of the stomach.
F. D. G. M. All the space from F. D. J. and M. down to the bladder is filled by the small intestines.
I. G. Last end of small intestine.
J. Jegunum, small intestine.
P. Pleura.
P. O. Lung.
R. Spleen.
R. D. Right Kidney, } They are not seen but are located
R. G. Left Kidney, } in the back.
L. C. D. Space where fluid will first be found in pleurisy.
S. I. Iliac S. leading to rectum. In front of which is seen the superior part of the bladder.
10 C. Tenth right rib.

The chyme is mixed with the bile, the pancreatic juice and the secretions of the various glands imbedded in the mucous membrane. In the small intestine is where the separation of the nutritive principles of food, is affected; this constitutes *chylification.* The nutritive part is absorbed to pass into our blood, the residue moves on into the large intestine.

§III. LARGE INTESTINE.

The large intestine is short, only about five feet long. Its position needs no description as plate III will show it very well.

The large intestine absorbs very little and takes but an inactive part in digestion; it serves more as a reservoir for that portion of the food intended by nature to be thrown away.

§IV. DIGESTIVE JUICES.

There are four digestive juices: the saliva, the gastric and the pancreatic juices, and the bile.

§V. SALIVA.

The saliva begins digestion, in the mouth, by penetrating every particle of food, and its action goes on even in the stomach, helping the gastric juice.

§VI. GASTRIC JUICE.

The gastric juice and saliva transform, in the stomach, all the albuminoid principles into *pepton*, which passes directly, it is claimed, into the circulation to form the albumen of the blood.

§VII. PANCREATIC JUICE.

The pancreatic juice is so rich with albumen that it coagulates by heat. It is absolutely essential for the minute division of the fatty vegetable and animal principles; it is only in this extremely divided state they can be absorbed and assimilated. The pancreatic juice is secreted by the *pancreas* (or pancreatic gland), located near the stomach as seen in plate IV.

§VIII. BILE.

The bile is a yellow fluid secreted by the liver, and is thrown into the small intestine (duodenum) through a canal, the *choledoch duct*, at the hours of digestion *only*. If there is too much bile secreted the surplus goes to the reservoir, called *gall bladder* (plate IV), through the cystic canal.

In the healthful state there is never any bile in the stomach, if any is found it means that there is a trouble in the stomach or that an emetic has been administered. The bile is the antiseptic of the bowels, it keeps that part of the food which is to be thrown away from putrifying in the bowels.

CHAPTER III.

§I. DIGESTION.

Digestion is the most important function of our organism: Its influence over body and mind is incontrovertible. An easy digestion makes one feel in high spirits and cheerful;—the limbs feel heavy, the brain is less active and the temper is none too good when digestion is laborious.

The stomach and the intestines, weakened either by excess in eating or drinking, or by any other cause, first will not work well, and if the excesses recur too often, the stomach will become irritated, finally contracting one of those horrible diseases called, neuroses of the stomach, gastralgia, chronic gastritis, flatulent dyspepsia, etc., etc. Pity to those who do not promptly put themselves under the care of a physician, for they will ever be miserable and in constant pain.

§II. WHERE FOOD IS DIGESTED.

Physiology teaches us, contrary to what is generally believed, that all food is *not* digested in the stomach.

The meats and the azotous part in eggs, milk,

cheese, and vegetables are digested in the stomach or rather dissolved by the gastric juice.

Fats, oils and fecula in the food are not digested or transformed in the stomach; but when the chyme passes into the small intestine (duodenum), where it meets the bile and pancreatic juice, by the combined action of which these fats, oils and fecula are divided into extremely small particles, and absorbed by the abdominal veins to pass into our circulation to be deposited in our organs.

The different salts contained in food or water are dissolved in the stomach and pass to our body, *i. e.*: chloride of sodium goes specially to the cartilages; —the phosphates and carbonates of lime go to the bones;—the salts of potassium, magnesium and some sodium to our muscles;—the salts of iron to our blood;—sulphur, silica and iron to the marrow and to the hair.

CHAPTER IV.

CLASSIFICATION OF FOOD.*

Food is divided into three classes:

1st. *The Hydro-Carbonates* composed of water and carbon, such as sugars, gums, fecula, mucilages, starch, honey, etc., etc. These substances, part of our daily meals, give rise to the following phenomenon; "they are immediately destroyed by the general combustion which keeps up life, and, notwithstanding the enormous amount taken every day, chemical analysis will detect but very slight traces of them in our organs."

To this class also belong the vegetables, though some contain little or no azote; some, like our peas, beans, lentils, corn, etc., contain a certain amount.

2d. *The Albuminoids* composed of hydrogen, oxygen, of little carbon and of much azote; the flesh, blood and the cartilages of animals, gelatine, the gluten of corn, the legumine in beans, lentils, peas, etc.; the albumen in potatoes, etc., etc.

* There is a table in Chapter 15 dividing food into three classes: 1st, easy of digestion; 2d, moderately digestible; 3d, hard to digest. AUTHOR.

These foods furnish their substance to all our tissues, from our skin to the marrow of our bones.

3d. *The Fats* composed of much carbon and hydrogen, little oxygen and of no azote: tallow, animal and vegetable fats, oils, butter, etc., etc.

A. The food of the first class gives its carbon, which, burning unceasingly, distributes the vital heat to our body. These foods will suit nervous, excitable, delicate persons who, by their great activity, lose considerable nervous strength, the food of this class will have the advantage of keeping up their strength without excessive excitation.

B. The food of the second class is more suitable to the formation, to the preservation and to the growth of our organs, and especially to the nutrition of our muscular system. It will add to the muscular strength and increase it greatly. Of this class ought to be the food taken by convalescents with proper graduation as to quantity, etc.

C. All food of the third class is deposited in our tissues in the form of fat.

When the food of the first class, that is, that which feeds the untiring combustion, is suppressed or not well assimilated, form any cause, the fat of our body is reabsorbed and burnt to keep up our vital heat. Fat is to our living body what oil is to a burning lamp. That is why a fleshy (fat) person can go through a long period of absolute fasting, to which would succumb a thin person. We know that *Hibernant* animals will be three or four months without eating and will live on the fat accumulated during the preceding months.

Experience teaches us that :—When fats are excluded from our diet, or not assimilated, our body grows thin; whilst their excessive use will cause the disagreeable condition called obesity. Therefore fat and sedentary persons will do well to eat food mostly of the second class. Active, lean persons ought to eat food mostly of the other two classes.

CHAPTER V.

§I. MEATS.

I will say right here that cook-books will tell you how to cook both vegetables and meats; but, as a rule all foods that can be cooked *in* or *with their juice* ought to be so prepared. The best book on the art of cooking and serving every inmmaginable food is that of Mr. Ranhofer, the French chef of Delmonico's.

Meat (flesh of animals or of fishes) being the most important article of food, we will give a description of the various kinds and their special properties, with a few words as to which constitution will be more benefited by their individual use.

§II. DARK RED COLORED MEATS.

These meats are exciting and very nutritive because they contain so much fibrin and osmazone.* They increase the vital energy and the muscular strength. A long continued use of them is too exciting, and will cause the blood to become too

---

*Composed of creatine, creatinine, sarcosine, etc. First extracted by Thenard. AUTHOR.

plastic, too thick;* only strong stomachs can digest these meats. Weak, nervous persons with an irritable stomach will eat very sparingly of the following meats; stag, venison, hare, wild boar, pork, sheep of two or more years, horse flesh, etc. This latter one is a much better food than one would think, a well cooked steak of young horse is really good, as I happen to know from experience; but it makes a rather thin broth!

Of the birds you have wood-doves (queest), turtledoves, lap-wings, sparrows, woodcocks, snipe, wild geese and ducks, moorhens, plovers, widgeons, etc., and most all the wild birds.

The constant use of this class of food will cause intestinal irritation; additional to this, to plethoric persons is the danger of a stroke of apoplexy, as these foods thicken the blood.

§III. RED FLESH MEATS.

These are less exciting than the preceding, and can be taken, with generally good results, by all constitutions and in all seasons. Eaten with fecula, with vegetables, according to the individual taste, they compose the best of nourishments.

In this class are beef, mutton, pork of four to ten months old, etc., pigeon, lark, bustard, par-

*All acids or acid fruits as vinegar, lemon juice have the effect of making the blood thin. AUTHOR.

tridge, etc; fish of rose colored flesh as salmon, shad, lobster, tunny, etc.

As *a rule all meats* are more nourishing when roasted and ought to be cooked that way.

These meats form the most favorable nourishment for the building of the muscular system, especially when they are roasted.

The *fibrine* in the meat forms the muscular fibre.

§IV. WHITE FLESH MEATS.

In this class we find veal, lamb, kid, rabbit, young pheasant, quail, young partridge, chicken, rail, frog, oysters, halibut, etc., and all the white flesh fishes.

These meats are *not* so nourishing as those of the first two classes, but they are particularly called for when the stomach is weak. Convalescents, delicate, irritable, nervous women, and those whose constitution is bilious will do well to diet mostly on those meats.

CHAPTER VI.

§1. EGGS.

The most nutritious food after meat is undoubtedly the egg, which is, with milk, what is called a complete food.* The egg is composed of three parts: 1st, the shell; 2d, the white; and 3d, the yolk. The shell is not eaten. The white is composed of albumine and water mostly, of salts, sugar and of lactic acid.

The yolk of

Albumine Vitelline Fibrine	Azotous substances.
Oleine Margarine	Carbonated substances.
Salts, ammonia and fatty substance.	

The eggs mostly used are those of hen, duck, turkey, etc. Hen and goose eggs, in the order named, are the ones most exclusively used.

---

*We call a food complete, when it is composed of albuminous matter, of fat and of different salts which are found in the blood.

To be a complete nourishment however, the egg must be fresh and not cooked hard, as little of it will be dissolved by the gastric juice if hard.

Here is a good way to take eggs: Beat the whites of two eggs in some good spring water and drink this twice a day or oftener; twice or thrice a week include a yolk with the whites of the eggs. Almost any stomach will retain this very nourishing food.

A fresh egg will go to the bottom of the water, whilst eggs 15 or 20 days old will float on the water or midway between the surface of the water and the bottom of the receptacle.

§II. MILK.

Milk is composed of butyrum, four parts, caseum, three parts, sugar of milk, five parts, salts and water, 88 parts. Though milk is a complete food and that the human body can live on it in childhood, afterwards our body must have a mixed nutrition.

Milk is adapted particularly to nervous constitutions, to convalescents, etc., *but many weak stomachs will not digest it.* This is when "pepsin" is not to be found in sufficient quantity in the stomach.

I have known dyspeptics to live for months almost exclusively on hot or warm milk, either pure or taken with chocolate, where cold milk could not be retained by the stomach. It brings calm to the nerves, and it will make lean persons grow fat if they take it in large quantities. Boiled milk, with a little salt added to make it palatable, is one of the most healthful drinks. It soothes an irritated stomach, it nourishes the flesh tissues and tends to make the complexion clear.

§III. BUTTER, CHEESE, FATS AND OILS.

Butter is a food used the world over; it is more easily digested when a little salt has been added to it. Melted butter is always hard to digest.

Cheese ought to be eaten only when fresh; never when it has any odor, much less when *travelers* creep in it.

Fats and Oils. Those especially who want to add to their adipose tissue ought to eat fats and oils; they are easily digested, and are very nourishing when taken in small quantities with other victuals; they will be more easily digested if a little salt is added to them.

§IV. SALT.

Salt, or chloride of sodium, is a necessity in our nourishment, not only because it renders food more easily digested, but also because it furnishes to our blood the chloride of sodium it loses through the excretions: urine, perspiration, tears, etc.

Salt is one of the best digestives of fatty and albuminous substances; it promotes the flow of the gastric juice and is of great help in the work of digestion. Dose of salt 10 to 20 grains.

It is said that bathing the scalp daily with hot salt water will restore the HAIR to its NATURAL COLOR, if it is GRAY. It encourages the growth of hair and prevents baldness.

Sore eyes. Apply hot salt water of strength to suit sensation, morning and evening. Sopping and bathing the eyes therewith and then wiping them off dry, has a wonderful curative effect.

Heart burn, water brash. Common salt, adult, one-half teaspoonful in a little cold water, taken as occasion may require, gives instant relief.

I have known cases of dyspepsia to be cured by taking one-half a teaspoonful of salt in a glass of cold water every morning before breakfast; this also will restore LOST TASTE. The same being taken twice a day has cured BRONCHITIS.

Bee sting, spider bite. Apply salt and soda, mixed in equal parts; it affords instant relief.

Ulcers or suppurating wounds. Bathe in a solution of salt in hot water, this will prove a success.

Corns. Rub up some fine salt with beef tallow, pare away the hard part of the corn, and apply the mixture every night; in a week, no corn.

Bottles filled with brine make first-class homemade fire grenades. SALT IS DEATH TO BED-BUGS; sprinkle their haunts with salt and that is the end of them. Cod-liver oil may be rendered palatable by being salted, etc.

§V. BREAD.

Bread is a food prepared with flour and water which is made to ferment some by adding yeast.

Every vegetable substance, containing gluten, sugar and fecula, can be used to make bread with; flour of wheat is the best as it contains the most gluten, a substance which gives the property to raise to the dough; this fact renders itmore easily digestible. Home-made bread is the best to eat, or graham bread.

§VI. RICE, MACARONI, VERMICELLI.

Rice is a very nourishing food and an emolient at the same time; it is easily digested and can be taken by any weak stomach. Rice cake is a very good way to take it; also in soup.

Macaroni and vermicelli are more nutritive than bread and are easily digested. It is better to boil them in water before they are prepared with milk, this is because when made in summer they are a little acid from fermentation.

CHAPTER VII.

VEGETABLES.

To give a description of all the vegetables known is not to be thought of, only the principal ones will be mentioned, such as potatoes, peas, carrots, cabbage, beet, jerusalem artichoke, truffle, beans, corn, lentils, etc.

§1. POTATO.

Potato is one of the most important of cultivated plants, and is in universal cultivation in the temperate parts of the globe. It is undoubtedly the most universally used vegetable, rich and poor have it on their table; and if properly cooked the tuber is a nice and easily digested substantial food. Starch is extracted mostly from potatoes. Mashed potatoes are not easily digested.

The potato is an American indigenous plant, and when first taken to Europe, by the Spaniards in the beginning of the 16th century, it was cultivated in gardens as a curiosity and not for a general use as an article of food.

The tuber is the only part used and is composed of 21 per cent of starch, 74 per cent of water, and the other five per cent, of salts of potassium and sodium, of fats and sweet principles.

§II. BATATA OR SWEET POTATO.

The sweet potato is also an indigenous fruit of America. It contains a large proportion of sugar and water; it nourishes more than the ordinary tubercule, and furnishes an excellent and very digestible food.

A cake made with the scraped pulp of sweet potatoes and eggs is far more nourishing and superior looking than the ordinary potato and rice cake.

§III. JERUSALEM ARTICHOKE.

Jerusalem artichoke is a South America tubercule, not known enough here; it is a delicious food, containing sugar, albumen, water, salts, etc., and is one of the most nourishing vegetables; any weak and irritable stomach can digest it without trouble.

§IV. CARROTS, CABBAGES, BEETS, TURNIPS, ONIONS.

Carrots are not easy to digest; the fibres in them are not digested at all so that dyspeptics or weak stomachs will let them alone.

A diet of carrots is said to keep the skin clear, and if there are any discolorations they will be removed. Take them stewed.

Cabbage. The digestability of cabbage varies according to the way in which it is taken, raw or boiled; thus raw cabbage alone is digested in two and one-half hours; raw cabbage with vinegar, in two hours; and boiled cabbage takes four and one-half hours. It has a great tendency to cause either flatulency or acidity in the stomach; cabbage soup is not to be eaten.

Beet is native of the temperate part of Europe. It is moderately digestible and contains quite an amount of sugar. The pulp will make fat for those whose stomach can digest it.

Turnip is a native of Europe and the temperate parts of Asia. Though somewhat flatulent it will be rather easily digested; one must be careful not to eat it too often.

Onion. India and Egypt have cultivated it from the most remote antiquity; the young leaves or seedlings are used also, but the bulb is the principal part used. It is moderately digestible. A sprig of parsley eaten after one has indulged in onions will remove all traces of the vegetable.

§V. PEAS, BEANS, CORN.

Peas. Dried peas are a very wholesome food

either in soup, or boiled like rice and served with a piece of boiled meat. French peas are not so nourishing as dry peas, they contain so much water; they are flatulent as the skin is not digested, their use must be moderate.

Beans contain only two and one-half per cent of undigestible substances, and the other 97 1-2 per cent of substances are taken into the human system and assimilated. They are very nutritious; Boston baked beans are known the world over and are really delicious; they are apt to generate gases in the bowels.

Corn is said to be indigenous of Mexico; though grains of corn have been found in the cellars of ancient houses in Athens; also in the "Bibliotheque Nationale" (royal library), in Paris, has a description of corn been found in an ancient Chinese book; it is also spoken of in the Scripture, the word "corn" being found there; but it seems to have been neglected until the discovery of America, when its cultivation spread in the old world.

Corn, or maize, when well ground and given to convalescents in the form of corn meal, is one of the very best of nutritious foods; it is also highly recommended to those troubled with a chronic disease of the bowels or stomach. Ground coarse and roasted, corn will make a very pleasant and wholesome coffee.

§VI. LENTILS, SQUASH, TOMATOES, OLIVES, CUCUMBERS, CELERY, SALADS.

Lentils to be good and easily digested must be decorticated; taken in soup they are very nutritive. When *not* decorticated they will often cause flatulency.

Squash is a very important article of food for both man and animal; its pulp is an emolient, it is also a very substantial food if properly cooked.

Tomatoes are indigenous to the tropical part of America, it is also called "love apple". Some persons will digest them better picked ripe; they are mostly used in making up catsup, preserves, pickles, etc.; they appear with almost every dish in Italy.

Olives furnish an oil much used in medicine, and in the preparation of salads. Pickled olives are a delicious fruit.

Cucumbers composed mostly of water are a delicacy in cold climates, some stomachs cannot digest them.

Celery contains sugar, mucilage, starch, etc.; therefore it is a good vegetable to eat; the threads are not digestible.

Salads are not easily digested by the stomach; their seasoning will make the work easier for the stomach and intestines, but they usually take three to five hours to be digested.

CHAPTER VIII.

PASTRY.

To give a description of all kinds of pastry would require two or three volumes as large as this one; so that only general rules about pastry will be laid down here, and, if followed, will prove of great benefit to all.

The more fat used in a dough, the more indigestible it is; the same if much butter is used.

The more the dough raises the lighter and more easily digested it is.

The dough when not thoroughly baked is very flatulent and will stay an extremely long time in the stomach before the gastric juice can act upon it. Children and weak stomachs are better without pastry; only these lighter ones are they permitted to eat, rice-cakes, powdered or ground cakes, scraped sweet potato cakes, semolina cakes, etc. These are easily digested so are also sweet cakes of the "sponge" variety.

CHAPTER IX.

FRUITS.

*Pickles** are beneficial to a *logy* stomach as the flow of the gastric juice is favored by their use; but their immoderate use is quite often the cause of dyspepsia.

Melons of the "cantaloupe" group are not to be eaten by weak, nervous, irritable persons, as they are rather indigestible. When taken, a glass of wine such as Marsala, Oporto, Madeira, Sherry, Malaga, etc., ought always go with the melon.

Water-melons are very refreshing when one is thirsty but one must not eat much of them as they have a debilitating effect upon the intestinal tract.

Dates come from Asia and Africa. They are healthy, nourishing and emolient. They are very soothing for a cold in the chest.

Figs have the same qualities as dates; a syrup of figs has been made which is a mild laxative. They were known by Hercules and Hypocrates who recommended them. Plato speaks highly of the nourishing properties of figs. Turkey figs are

---

*Pickles are not clasified so are included here as fruit.
AUTHOR.

the best; they contain about 24 per cent of sugar-fig.

Dry Raisins and *Prunes* are very nutritious and are easily digested. Stewed prunes, before going to bed, are one of the best means of keeping the bowels regular. The author has always prescribed them successfully for constipation in women.

Peaches are native of Persia and of northern India. They are refreshing, nutritive and aperient. One can eat almost any number of peaches without fear. A little Malaga or Sherry wine added will make them a very pleasant dessert.

Apricots are less nourishing than the peaches, but are healthy and wholesome.

Grapes are very healthy; the seeds and skin *must not* be swallowed; hundreds of cases of appendicitis have been caused by seed of fruit lodging in the appendix (blind bowel) causing inflammation with death as a result. *Never swallow the seed of a fruit.* The best grapes are the Concord, Catawba, Malaga, Delaware, Tokay, etc.

Apples. Those somewhat acid and juicy are the best on account of the phosphorus acid in them; they act as a general tonic. Farinaceous apples are dry and lay heavy in the stomach; they ought to be eaten only when baked.

Pears have the same properties as apples.

The less nourishing fruits are: *Oranges*, *Blackberries*, *Strawberries*, *Blueberries*, *Plums*, *Cherries* of all kinds, except that called "choke cherries", etc.*

§II. HONEY.

Honey was considered by the ancient writers, as insuring a long and healthful life if it was used constantly; and we find Democritus, Hypocrates, Pythagorus, C. Asinius Pollio, Pringle, Cornaro† and others who constantly ate honey lived to be almost one hundred years old. Large quantities can be eaten even by very weak stomachs, and it will increase bodily weight in a very short time.

---

*Plums, peaches, lemons and similar small fruits keep best in papers. It will repay the housewife to do her perishable fruits up in paper as soon as purchased.

†Cornaro's constitution, naturally not strong, was greatly injured by intemperate eating and drinking, with other excesses; so that, when 40 years of age, he appeared to have little hope of prolonged life. At this time he adopted strict rules of temperance, which, co-operating with his general care of health, the constant use of honey, served to extend his life to nearly 100 years. His old age was remarkably cheerful and he ascribed it, also his good health, chiefly to honey.

CHAPTER X.

BEVERAGES. §I. WATER.

The most natural beverage is good running spring water and it is also the most generally used. But, according to the products of the climate he lives in, man will drink spirituous liquors made with any of the plants or fruits containing sugar.

Without water no organism can live; it has different effects according to its temperature.—*Cold*, it is an energetic tonic, but in summer it is dangerous especially when the body is in perspiration. —*Lukewarm*,it is emolient, laxative and will cause vomiting.—*Warm* water is exciting, sodorific and activates digestion, especially if it is taken with an equal volumn of rum, whiskey, kirsch, etc.; it is then called "Punch." Water with a little sugar has great dissolving power on the food in the stomach, but it must be drunk very slowly and *after* eating.

§II.

I will briefly state the different liquors which

each country makes from some plant, seed or fruit containing sugar.

England makes ales, porters, gins, etc.; Holland gins are very highly recommended; Spain, Italy and especially France furnish us with wines; France has always produced the best wines in the world. Persians have *cocanar*, made with salep, —the *cachiri* of Cayenne is made from powdered manioc;—the Indians make with corn a liquor called *chicona*;—the *saki* of the Japanese, and the *arak* of the Arabs are made with fermented rice; —the negro of Africa and the natives of Australia make a wine with the milk of the *cocoanut*:— *Rum* is made in the South and in Jamaica, in distilling the sugar of *sugar cane*; Poles have their *lipet* made with honey, etc.

§III. WINES.

The best wines come from France, but Italy, Spain, Germany and now the United States furnish also very good wines. Mr. Nicolas Longworth of Cincinnati was the first American to make American champagne, in 1850, after some difficulties. From 1850 to 1895, the wine industry has grown stupendously in New York state, along the Hudson, in the valley of the Ohio river, in California, etc., until now we can say that the United States are making a good part of the wines drunk by the Americans.

Here are the different kinds of wines which great men preferred:—Napoleon Bonaparte was particularly fond of the Chambertin (a rich Burgundy wine),—Fredrick the Great prized the Tokay more than any other,—Richelieu drank but the wine of Romance,—Peter the Great only of Madeira,—Rubens that of Marsala,—Cromwell the Puritan adored Malmsey wine,—Balzac's glass was full either of Champagne or of wine of Vauvray,—Goethe took Johannisburg,—Humbolt had Sauterne,—Lord Byron was especially fond of Port wine,—Francis 1st of Sherry wine and Henry IV of the wine of Suresne, etc.

First class. The following wines are of great value to weak, lazy stomachs to activate digestion of food; they act by the alcohol and tannin in them, they are: The wines of Bordeaux, of Burgundy, of Champagne of Sauterne, etc. These departments* give these particular brands (which are the best in their class): The Volnay, the Beaune, the Nuits, the Pomard, the Chambertin, the Closvougeot, etc.

Second class. The wines of this class are those containing a large proportion of alcohol, 15 to 25 per cent, and will quickly act upon the head and legs of the unsuspicious drinkers. They must be taken in small quantity and diluted with water.

---

*These are departments or provinces in France.
The AUTHOR.

The wines of Marsala, of Oporto, of Madeira; the strong wines of Languedoc, of Provence, of Roussillon, and of southern Europe, are the best, and ought to be used very moderately.

Third Class. This class is composed of the sweet and generous wines, such as: the Samos, the Cyprus and all the Greek wines; the Malvoisie, the Lunel, the Frontignan, etc. These wines are drunk at dessert and are *sipped.*

Fourth Class. The Champagne, the Grave, the Tokay, the Limoux, the Rhine wines, etc., contain but little alcohol, they are easily digested and are exhilarating.

§IV.

Here is a table of wines with their per cent of alcohol:—

Lissa,	24.69
Marsala,	23.83
Madeira red,	20.52
Madeira white,	20.00
Oporto, (or Port)	20.00
Constance white,	18.17
Teneriffe,	20.00
Lacryma Christi,	19.50
Malaga,	17.42
Bagnols,	17.00
Ermitage,	17.00
Roussillon,	16.88
Malvoisie,	16.50
Johanisburg,	15.16
Malaga ordinary,	15.00
Sauterne,	15.00

Cyprus,	15.00
Burgundy,	14.70
Rivesaltes,	14.60
Jurancon red,	13.70
Lunel,	13.70
Angers,	12.90
Champagne,	12.77
Grave,	12.30
Rhine wines,	12.50
Beaune white,	11.80
Frontignan,	11.80
Cote-Rotie,	11.30
Macon white,	11.00
Volnay,	11.00
Orleans,	10.66
Bordeaux red,	10.10
Saint Emilion,	10.00
Vauvray white,	9.66
Chateau-Latour,	9.33
Chateau Margau,	8.75
Chateau Laffite,	8.75
Chablis white,	8.00

§V. ALCOHOLIC LIQUORS.

The name strong or hard liquors is given to brandy, fine champagne, gin, rum, whiskey, rye, kirsch, arak, etc., because they contain a large percentage of alcohol and are very intoxicating. These liquors taken, while eating or after, in small quantity, will help digestion especially when taken in a little sweetened water (punch).

§VI. MALT LIQUORS.

These liquors are agreeable, healthy and nourishing; they will add greatly to one's adipose tissue; lean persons will be greatly benefitted by a good ale or porter, while stout persons ought not to touch them. They contain but little alcohol, the ales or light colored beers contain from 2 to 4 1-2 per cent; the porters or high colored beers contain from 5 to 8 per cent. One cannot be too careful in the choice of malt liquors, as unscrupulous manufacturers will adulterate them by adding bitter extracts to give them more zest and taste.

§VII. CIDER.

Cider is made mostly with apples and contains from 2 to 4 per cent of alcohol; fresh cider is good but if taken often or immoderately it will cause diarrhœa and even dysentery. Cider has a bad effect upon the teeth. Travelers through France, in the departments of Picardy and Normandy, are pained to see pretty young girls, real pictures of health, the smile of which will show a set of decayed teeth or no teeth at all. So when one drinks cider the teeth must be thoroughly washed immediately after.

Cider of pears called *perry* is slightly more alcoholic than that made with apples, consequently is to be taken moderately; it is a very good beverage.

§VIII. HYDROMEL.

This beverage was made by the Romans, the

Greeks and even by the Egyptians. Pliny, the naturalist, has even transmitted the formula: honey, one part, rain water, three parts; heat gently and take the froth away carefully. You will know when the hydromel is done by dropping a *whole* fresh egg in, if it stays about half way between the surface and the bottom, take the liquid off and let cool, then put in a small cask, placing over the bung-hole a small board which will let the froth made by fermentation escape; warm water must fill the space left by the escape of the froth; when fermentation is over close tightly the bung-hole and put the cask or keg in the cellar *on the ground.* After three or four months pass it through a filtering paper, then put in bottles.

A little yeast dissolved in water might have been added, so that fermentation would not have lasted more than three or four weeks.

§IX. CORDIALS.

Cordials like Chartreuse, Benedictine, Curocoa, Trappistine, etc., are taken at dessert. They are highly aromatic and ought to be taken only on rare occasions.

§X. CONSIDERATIONS.

Alcoholic liquors are very important, even absolutely necessary in a great many cases of sickness; also when the stomach is weak they will stimulate it. But with some cases of dyspepsia, liquors will do great harm.

Alcohol is for the most part rapidly absorbed in an unchanged state, either in the form of liquid or vapor; and this absorption may take place through the cellular (or connective) tissue, the serous cavities, the lungs or the digestive canal. It is, however, only with absorption from the intestinal canal that we have to deal, in relation to man. Almost the whole of this absorption is affected in the stomach, and it is only when alcohol is taken to great excess or is mixed with a good deal of sugar that any absorption beyond the stomach occurs. The rapidity of absorption varies according to circumstances. It seems to be proved beyond all doubt that "alcohol stays for a time in the blood, that it exercises a direct and primary action on the nervous centers, whose function it modifies, perverts, or abolishes, according to the dose, that neither in the blood or in the expired air are any traces to be found of its transformation or destruction, that it accumulates in the nervous centers and in the liver, and that it is finally discharged from the system by the ordinary chanels of elimination." The alcohol, when it has entered the blood, is diffused over the organism, remains during apparently different periods in different organs, and almost immediately begins to escape. The fact of the retention and accumulation in the nervous centers and liver tends to throw much light on the special diseases of drunkards. The action of every kind of alcoholic drink in moderate doses is that of a somewhat rapid stimulant. The bodily and mental powers are for a time excited beyond their ordinary strength, after which there is a corresponding depression. Although the alcohol which is introduced into the system cannot act as a true food, it indirectly takes the place of food by diminishing the wear and tear of the system and thus rendering less food sufficient; a fact which is proved by chemical experiments showing that less carbonic acid and urea are given off when alcohol is administered in moderation than when it is totally withheld.

But great care must be taken lest one might fall into intemperance as the road is agreeable and gay; it is very easy to become a drunkard, and that is the greatest curse a man can bring upon himself, family and society.

The only places where intemperance can successfully be fought are in the schools and in the family. Let us teach our children (as the Spartans did theirs) to hate intemperance, and show them how disgraceful a man is when drunk. Man must be taken when a child, when he is in such a state of moral *receptivity* that a lasting impression can be made upon his mind; teach him his future obligations to himself, to society, to his family, to his country; fill him with horror for the drunkard and liquor, and gradually that curse called intemperance will be eradicated from our midst. Such teachings will be worth more than all the temperance lectures in the world, for those attending such lectures are "temperance people," and you will not find a toper there to listen to the speeches.

§XI. VINEGAR.

Vinegar is the result of the transformation of spiritous liquors by acid fermentation. Vinegar, in small quantity, has the property of promoting the flow of the gastric juice, of dissolving albuminous substances and of making digestion easy.

Vinegar and all acids have the effect of thinning the blood, and diminishing the volume of its globules. The continued use of acid foods or liquids will ruin the stomach and the strongest constitution. Young ladies with an undue tendency to corpulency sometimes drink vinegar freely, with the view of improving the figure, but as vinegar only causes thinness by injuring the digestion, it is obviously not worth while that they should run the risk of exchanging a slight fullness of habit for chronic dyspepsia.

§XII. SUGARS.

The principal ones are cane, beet, and maple

sugars; we also have fig, date, and milk sugars, but they are less important.

The first three are the most important and the most used. Every day they enter the preparation of our food.

They will help to form the adipose tissue of our body, and they will also supply the carbon for our respiration. Sugars are a great help to the stomach, as, under their influence, food will be quickly digested. Quite a quantity of sugar can be eaten without injury, avoiding excess however.

CHAPTER XII.

§1. COFFEE.

Coffee is a native of Abyssinia and Arabia. Leonard Ranwolf, a German physician, was probably the first to make coffee known in Europe, by the accounts of his travels printed in 1573.

The beans should never be darker than a light brown color, which is quite sufficient to bring out the excellent aroma and other qualities of the coffee; and when the roasting is carried further, more or less charring is the result, and a disagreeable burned smell is produced, which tends to overcome the natural pleasant aroma.

Coffee is a stimulant and will not injure the stomach or bowels unless taken immoderately. Its principal action is upon the brain, during that special excitation, the ideas are clearer, more concise, more gay, the imagination is bright and active, in short a well prepared cup of Moka taken before sitting down to study or write is a great help to the brain.

Coffee does not retard the action of the bowels, as strong infusions of tea tend to do, partly because there is less of the astringent principle, and also owing to the presence of the aromatic oil which tends to move the bowels. The important offices which coffee fulfils, are: to allay the sensation of hunger, to produce an exhilarating and refreshing effect, and most important of all, to diminish the amount of wear and tear, or waste of the animal frame, which proceeds more or less at every moment.

Irritable or excitable persons will do well to take coffee only at long intervals, as it might make them still more nervous and excitable. Voltaire, who lived to be eighty-five years old, took enormous quantities of coffee. Fontenelle, who also drank coffee all his life, died at 90, so that the best guide to see if coffee will injure, is one's physician; he ought to be consulted much oftener than he is, about one's food or beverage.

Coffee, to be good, must be made in a coffee-pot which will give boiling water time only to dissolve the fatty and azotous principles, the soluble salts and aromatic essence; the cellulose, the tannin and the lime salts ought to be left in the grounds. Here is a simple way to make a cup of good coffee. Take one teaspoonful of ground coffee for each cup to be made, put it into a strainer, then pour boiling water over it; the water is to be boiled and poured over three times in all; this will give a very nicely flavored cup of coffee; not more than two cups are to be prepared in this way at one time.

The best coffees are the Java and the Mocha, the Sumatra, Brazil, etc. Mexican coffee is said to be very good.

§II. TEA.

Tea is said to have been introduced into China itself from Corea about the fourth century of the Christian era, and to have extended to Japan about the ninth century. There are two kinds of tea, the green and the black teas ; they come from the same tree, but the leaves have been harvested at different degrees of maturity, and also have been roasted differently. The green contains more essence than the black tea, and is consequently more stimulating and exciting. The *real* gunpowder tea is *never* exported from China, the imperial family has the crop and death would be the penalty to those who would sell this tea to any one. The gunpowder tea we have is an imitation of the imperial tea (Chou-Cha).

While some physicians have over-praised its value, others have regarded it as a source of numerous diseases, especially of the nervous system. Old and infirm persons usually derive more benefit and personal comfort from tea than from any other corresponding beverage. Tea, in the form of a cold, weak infusion, is often of great service in fevers. For persons of a gouty and rheumatic tendency, and especially for such as are of the *Lithic Acid*

Diathesis, week tea, taken *without* sugar, and with very little or no milk, is the best form of ordinary drink. In some forms of diseased heart, tea proves a useful sedative, while in other cases it is positively injurious; a cup of strong green tea, especially if taken *without* sugar or milk, will often remove a severe nervous headache. Tea will cause sleeplessness and it therefore is not to be taken at night.

§III. COCOA.

Cocoa is a native of the tropical parts of America. Cocoa is very nutritous. The principal constituent of cocoa beans is the soft, solid oil called *cocoa butter*, which forms more than 50 per cent. of the whole shell bean, about 22 per cent being starch, gum, mucilage, etc., and 17 per cent being gluten and albumen. The seeds are the only part used. For dietic use, cocoa is prepared in several ways. It is made into chocolate, and it is in that form that it is mostly taken.

Chocolate is made from the seed of cocoa, reduced to a fine paste, and mixed with powdered sugar and spices. The paste is then poured into moulds, in which it is allowed to cool and harden. Chocolate is used as a beverage, and for this purpose is dissolved in hot water or milk. The merest dash of cinnamon in a cup of chocolate after it is poured adds a piquant and indistinguishable flavor. It is especially adapted to delicate constitutions, to convalescents, to children, to old persons, to nervous, irritable, weak stomachs, and to all persons not able to digest abundant food. It is wholly digested as no traces of it can be found in the fæces.

CHAPTER XII.

GENERAL CONSIDERATIONS ABOUT FOOD AND TEMPERAMENTS.

§I. SANGUINE TEMPERAMENT.

What constitutes the sanguine temperament is the predominance of the circulatory and respiratory systems over the others of our body.

The best food for such a temperament is the different roast meats, vegetables, white flesh fishes, chicken, etc. ; pure water is the best drink, as any alcoholic liquor or beer will excite the system too much. Abstention of sugars, fats, farinaceous vegetables as much as possible, as these substances will cause one to grow stout very quickly, and will predispose those of a sanguine temperament to apoplexy.

§II. BILIOUS TEMPERAMENT.

In this instance the stomach and liver are the predominant organs of the body.

Fruits and drinks somewhat acid are very good for this temperament ; the regime must be composed of white meats, fresh fish, aqueous vegetables, etc.; no alcoholic drinks, nor too much spiced or irritant food if the bilious man values his health. He always ought to get up from the table feeling a little hungry, rather than to eat as long as his appetite will last.

Those with bilious temperament are especially liable to have "Flatulent Dyspepsia."

§III. NERVOUS TEMPERAMENT.

This temperament, as is well known, is due to a greater activity of the nervous system than of the other systems.

In this temperament the regime must be calmative and restoring; white meats roasted or broiled, white flesh fishes, fresh eggs raw or very slightly cooked; vegetables, vermicelli, rice, macaroni, fruits, etc. In short nourishing and laxative food. No fats, pork, goose or any other exciting meats or food.

The stomach, with those of a nervous temperament, is almost a tyrant, requiring constant care and attention, for the least error in the regime will be resented by the oversensitive stomach, and days, nay, years of suffering are the penalty. There are pains in the head and back, weariness in the limbs, etc.; some persons are gloomy and pessimistic, worry unnecessarily, others complain of weak memory, and inability to concentrate their thoughts;

others will suffer from vertigo; there may be palpation of the heart, sometimes very severe, etc. All these symptoms will be brought on by errors of regime or by excesses. What is said here is a deplorable truth. Question those who have nervous dyspepsia, and you will find their answer tallies with the above.

CHAPTER XIII.

OBESITY.

Obesity is too well known to need description; the treatment of this infirmity is the only part to interest us.

In ancient Greece, and especially with the Spartans, obesity was a dishonor, they said that a man could not be a warrior; and a woman could not take care of her children.

It is remembered that in the classification, *fat*, *gelatinous*, *feculent*, *farinaceous* foods, beverages such as beer, cider, sweet wines, make a chyle that will *in toto* be transformed into adipose tissue, therefore it is evident the corpulent person ought not to eat this class of food.

The regime to follow will be composed exclusively of dry stimulants, high seasoned foods: meats composed of much carbone and azote as beef, mutton, hare, venison, the fat of which having been thoroughly taken off; partridge, thrush, pheasant, pigeon, etc.; french peas and beans, spinach, endive with sugar, *never* with fats; all kinds of salads highly seasoned with vinegar, acid fruits, etc.

The beverages to drink are white wines well diluted in hot water, lemonades, seltzer water, black coffee, etc. Occasionally also ought to be taken sudorific and diuretic draughts, mild purgatives, to promote loss by the perspiration, by the urine and through the bowels ; but great care must be taken lest the intestinal tract might get out of order. A STOUT PERSON OUGHT TO DRINK AS LITTLE AS POSSIBLE, AND MOST ALWAYS WARM DRINKS.

The energy of this regime is increased by salt water ablutions all over the body ; and by frictions with hydriodate of potassium dissolved in alcohol, made once or twice a day over the parts where the most adipose tissue is, as over the chest and abdomen.

Hydriodate of Pot.,	5 ounces.
Alcohol,	2 lbs.

Wash with half a tumbler each time ; the frictions to last from a quarter to half of an hour.

Hunter says that a mild compression over the parts to be reduced is an excellent way of increasing the activity of the vessels absorbing the fat ; the abdominal belt has no other power to diminish the fat than through its mild compression.

But, to exercise enough to cause perspiration and fatigue ; to rise early ; to go to bed late and to give little time to sleep, five or six hours for instance, must of necessity, with the other means, cause a loss of adipose tissue and bring back the body and muscles to their normal state.

One other treatment that is exceedingly beneficial to those troubled with obesity is the electric spark. It is a "dissolvent of fat." It acts by increasing the activity of circulation, causing a rapid absorption of the globules of fat. Many ladies, without being really corpulent, are rather fleshy and puff vigorously when they walk. The electric spark will cure this if properly applied. This treatment is to be taken only under a physician who understands electricity thoroughly, given out of place it will injure the person treated.

The following case is very interesting:

A lady, Mrs. G——, about 35 years old, was corpulent, weighing some 180 pounds, and could not walk without being "all tired out" and "puffing vigorously"; she could not walk easily, her ankles were weak and the feet were tender, they would become very sore. After having tried a regime under different physicians, and after taking much medicine which upset her stomach, she came to me.

I first gave her a list of food and drinks to be taken the first week, then would change it at the end of every week for ten weeks. In the meantime I gave her two or three electric treatments every week. After two and one-half months she started for the "World's Fair" in Chicago, and was gone four weeks. This is what Mrs. G—— told me after her return:

"I walked from morning till night for four solid weeks, and I never felt so well, every morning I started as fresh as ever, would walk all day and get back to our hotel only at night. I never walked so much and with so little fatigue in all my life."

Many other treatments, more or less absurd, have been devised either violent or not; such as the ablation of the fat by operation, the use of violent, drastics, the absolute diet, etc. They are not to be thought of; other treatments, as drinking vinegar, are injurious. There is one thing about the treatment of obesity, as well as of any other ailment that is always to be kept in mind, *i. e.*: the danger of, and the impossibility of treating every person alike, each temperament, each constitution, each person even of the same temperament, will require a different treatment, so that no absolute rule can be laid down.

CHAPTER XIV.

LEANNESS.

I do not mean to say anything about "the growing lean" caused by sickness, such as typhoid fever, tuberculosis, cancer, etc., I mean to speak of that leanness compatible with health but not with good looks.

This infirmity may be caused by excessive irritability of temperament, or be due to the moral or physical state of the individual. Thus the nervous and melancholy temperament, the mournful passions, protracted work, excessive heat in summer, fasting, insufficiency of food, its bad quality or imperfect preparation, the abuse of liquors, of tobacco, etc., are causes of leanness. If it is caused by any sickness, this must first be cured. When the cause has been removed, you can then by the following regime build one's system up and increase the adipose tissue.

When leanness is not caused by an organic lesion, the first step is to stop whatever cause entertaining it, and to have the person follow a regime which will invigorate and lessen the wear and tear losses of the system, caused by the natural exchanges going on in our body.

For example, let a person eating highly seasoned food, drinking exciting beverages, dry wines, tea, coffee, etc., who spends a great deal in intellectual and physical activity ; let such á person change her mode of living, and make choice of *fatty*, *gelatinous*, *feculent* substances ; drink sweet wines, beer, porter, cider ; let her give little time to work and a good deal to sleep, in short, let her give much to animal life and very little to intellectual life. A few weeks of this regime must of necessity make one stout.

The food to take is : All kinds of fecula, of mealy foods, prepared with fats, fats and fat meats, milk, Jellies, consommes, oils, butter, fresh cheese, rice, salep, sago, chocolate, honey, dates, olives, cocoanut, vegetables prepared with much butter or fat, figs, grapes and sweet fruits. For beverage : Cider, beer, porter, sweet wines, Maderia wine, two glasses a day, hydromel, weak alcoholic liquors.

It is better to eat different kinds of food even at the same meal, as the stomach, if given the same food too often, will grow *logy*.* The regime also requires luke warm baths ; friction all over the body with a flannel either dry or wet with aromatic wine, two or three times a week ; a light laxative every week to clean the digestive canal of all saburra that may have formed.

---

*Sluggish, inactive, lazy stomach.

CHAPTER XV.

*THE SKIN.

It is not possible to deny that food has some action upon the skin ; all physicians and specialists are in perfect accord on that point. Food forms the blood and the blood all goes to the capillary of the skin, the impurities in the blood are deposited there, the consequence of which is bad complexion. A pretty complexion is a boon to everybody who is so fortunate as to have it, for it not only adds to the beauty, but it testifies to the good health of its possessor.

§1.

Food will have immediate or slow action upon skin.

First. Immediate action. When food reacts upon the skin a few hours after taking it, it is either after its passage into the circulation acting as medicine would, pathogenic action, or by creating trouble in the stomach and intestines, which trouble

*Pidoux, Rapport de l'herpetisme et des dyspepsies, Paris. Brocq. Traitement des maladies de la pean. Paris.

reacts over the whole system. In this last instance the way the food has been prepared, cooked and eaten is also an important factor.

The skin will, sometimes after a meal, be covered with nettle-rash or erythema, these eruptions cease as soon as the cause is removed.

Foods that may cause immediate eruptions on the skin are :—Sardines, salmon, mackerel, herring, gold fish, salt water fishes, oysters, mussles, lobsters, shrimps, crabs, cress, salt and smoked meats ; strong tea or coffee will sometimes make an itching much worse, if it does not exist it will often be brought on, alcoholic liquors, certain acid fruits, as oranges, lemons, etc.

But there is no invariable rule, as some persons will eat any of the above substances with no after effect on the skin ; while they will be unable to eat turkey or some other food without becoming covered with nettle-rash.

Illustrating this impossibility of any fixed rule is this one case. A child of about 12 has been unable to eat anything in which the yelk of egg is used, her mother discovered the fact by accident some years ago ; and without failure every time the child takes food with yelk of egg in, she is covered with nettle-rash the next day.

Some persons cannot eat fat or butter, others oils or highly seasoned food, etc.

Second. Slow action. (This food should not

be taken by gouty persons.) This class of food of slow action contains the following :—Coffee, alcoholic liquors, tomatoes, wines, dark or red colored meats, etc. (What should be given to gouty persons is : Milk, white flesh meats, cooked green vegetables, sherry wine, etc.) There is also salted cheese, canned meats, candies, bon-bons, pastries, bran, buckwheat, gruel, oat meal, etc., that are not to be eaten *too often*.

Of course these different kinds of food are not to be proscribed, but "*If one has a skin trouble it would be well to see if there is not one or two kinds of food that are somewhat often on the dinner table.*"

§II. HYGIENE OF THE SKIN.

This will be a very brief resume of the hygiene of the skin.

The skin is, as is well known, an organ of secretion, of excretion, and to a certain degree, of absorption.

Its perfect working is necessary to health and to life, as many celebrated examples have proved. Therefore every effort ought to be made to insure this perfect working of our cutaneous tissue. Perfect health is impossible without free perspiration. This is one of the methods adopted by nature to free the system of waste material, and one can see how necessary frequent bathing is to remove the impurities forced through to the pores of the skin.

The best way to obtain it is to always keep scrupulously clean, washing the whole body (twice a week or oftener in winter, every day in summer), rubbing well with a dry flannel after the bath. The water ought to be cool. Never take a hot bath (except under the advice of your physician), as it weakens the nervous system very much; a cold bath is far better for it has a tonic effect upon the skin and nerves. At first one must take luke-warm baths, then gradually each bath is taken with the water a little colder *until it can be taken with the water real cold.* The shock of a cold bath will naturally be followed in a few seconds by a strong reaction.

Soft and flabby skin will gain freshness of texture by the use of cold water to which has been added a little common salt. For an irritable skin add about one-half to two pounds of gelatine, of starch or of bran to the water. For a rough dry skin add about eight ounces to one pound of glycerine to the water.

Chapped hands are always the result of the extremes of hot and cold to which the tender skin at the back of the hands is subjected. In even temperature this cuticle, as sensitive almost as the eyelid, will not roughen and peel in an attempt to form a defence against the cold, and unless the proper applications are used one must be careful to protect these extremeties from sudden chill and heat. In winter two baths a day in tepid water with castile soap should be sufficient to keep them clean; and, when walking, the heaviest dog-skin gloves, cotton lined, afford the only proper protection. Muffs are a snare and delusion, unless the hands are held in the fur protector continuously, and when all precautions have failed here is an excellent remedy for restoring and softening the skin. It is a home made cold cream, as efficacious for lips as hands, and easily made by any one. One cake of white wax, a parafine candle, and a bottle of sweet oil. Prepare by cutting the candle away from its string and breaking into small pieces, and shaving the wax like soap, mix this with the sweet oil and let it melt and combine in an earthenware bowl set in a saucepan of hot water. When this grows boiling hot pour into it two tablespoonfuls of rice water, a teaspoonful of violet perfume or extract, and the same amount of extract of white rose. Now remove the saucepan from the stove, stir the compound rapidly until it begins to grow cold. This last must be a gradual process, for if the cream is subjected to a sudden chill it is apt to grow too hard and very lumpy. Hot vinegar also keeps the hands from chapping.

Carbonate of ammonia is to be used if the feet and arm pits are odorous and smell none too sweet; about two ounces to a pail of water; permanganate of potassium, 20 grains to a pail of water, is also used to advantage.

Those whose skin is oily will do well to put about three ounces of powdered borax in the water; or two to six ounces of carbonate of sodium.

When the secretions of the skin are odorous, use either tincture of benzoin, tincture of lavender, cologne or other aromatics in the water.

Sulphurous, vapor, etc. baths all have their usefulness, but ought not to be taken except under the advice of a physician.

Some women have preserved the freshness of the skin and firmness of their breast by bathing every morning their arms, shoulders and the breast with *cold boiled water*.

If the skin is too dry and scaly it ought to be rubbed now and then with *fresh* cold cream, or with glycerated starch; if it is oily, (besides the use of carbonate of sodium, or borax), once or twice a week, a good washing with tar soap or borated soap will keep it smooth and sweet; a few drops of cologne or of tincture of benzoin, or of lavender, are to be added to the water used.

The skin may be, when not as good as it should be, improved by care of the diet, by the proper use of the bath, and by avoiding whatever makes one ill or even dull and apathetic.

One of the most vexing of the enemies to a good complexion are black-heads or flesh-worms, which are very liable to degenerate into ugly looking white pimples, by no means endowed with a life of their own, although this is the general belief concerning them. They are often the result of uncleanliness, for the oily matter exuded by the pores form these black specks, which dishonor the prettiest face and are in reality nothing but accumulated dirt. When black-heads have once been permitted to form, it is quite a difficult matter to get rid of them. Many applications are recommended for the purpose ; for instance, bicarbonate of soda dissolved in hot water, borax water, white of egg applied to the skin on retiring to bed, pure alcohol, etc. ; one good way of getting permanently rid of black-heads is to thoroughly wash, night and morning, with water as warm as it can be borne, and then bathe the face for ten minutes at least in tepid milk by the aid of a soft and very fine sponge. Continue this for a month, and you will find that your skin has become pure and sweet as a baby's.

A diet of carrots is said to keep the skin clear, and if there are discolorations they will be removed. Take them stewed.

TABLE OF FOOD.

EASY OF DIGESTION.	MODERATELY DIGESTIBLE.	HARD TO DIGEST.	EMOLIENT AND LAXATIVE, AND EASILY DIGESTED.
Beef.	Mutton.	Pork.	Arrowroot.
Hare.	Lamb.	Veal.	Salep.
Sweetbread.	Venison.	Goose.	Prunes.
Chicken.	Rabbit.	Liver.	Manna.
Turkey.	Young pigeon.	Heart.	Vegetable jellies.
Partridge.	Duck.	Brain.	Rhubarb plant.
Pheasant.	Wild waterfowl.	Salt meat.	Tapioca.
Grouse.	Woodcock.	Sausage.	Sago.
Beef tea.	Snipe.	Mackerel.	Jujube.
Mutton broth.	Soups.	Eels.	Squash.
Milk.	Eggs.	Salmon.	Stewed apples.
White of eggs, raw.	Butter.	Herring.	Pears, (soft juicy pulp.)
Tarbot.	Turtle.	Halibut.	Figs.
Shad.	Cod.	Salt fish.	Grapes.
Flounder.	Pike.	Lobster.	Dates.
Sole.	Trout.	Crabs.	Peaches.
Fresh fish, generally.	Raw or stewed oysters.	Shrimps.	Raspberries.
Roasted oysters.	Potatoes.	Mussels.	Mulberries.
Stale bread.	Beets.	Cabbage, boiled.	Strawberries.

Rice.	Turnips.	Hard cooked eggs,	Blueberries.
Honey.	Cabbage, raw.	Fresh bread.	Honey.
Tapioca.	Spinach.	Oil.	Cherries of all kinds except that called choke cherries.
Sago.	Onion.	Butter, melted.	Boiled Onions.
Jerusalem artichoke.	Artichoke.	Cheese.	Spinach, called in France the "broom of the stomach."
Arrowroot.	Lettuce.	Muffins.	Ales.
Asparagus.	Oranges.	Buttered toast.	Porters.
Squash.	Cellery.	Pastry.	
French beans.	Apples.	Cakes.	
Cauliflower.	Apricots.	Custards.	
Lentils.	Strawberries.	Nuts.	
Sweet potatoes.	Raspberries.	Pears.	
Baked apples.	French bread.	Plums.	
Chocolate.	Farinaceous puddings	Pineapples.	
Grapes.	Jelly.	Cucumbers.	
Currants.	Rhubarb plant.	Onions.	
Peaches.	Cooked fruits.	Carrots.	
Toast-water.	Cocoa.	Parsnips.	
Tea, black.	Coffee.	Beans.	
Sherry.	Porter.	Mushrooms.	
Hydromel.		Pickles.	
Claret.			
Punch.			
Ale.			

PART II.

CHAPTER XVI.

DYSPEPSIA.

Dyspepsia is far from being a new disease, it was known to the very earliest writers, to Galens, Hypocrates, and others. Up to to-day many authors have written about dyspepsia as a disease in itself, but "*Dyspepsia is only the symptom of a troubled action of the stomach,*" nothing more. When one says "I have dyspepsia, he has a digestive disturbance that may be caused by cancer of the stomach, locomotor ataxia, tuberculosis of the stomach, catarrh of the stomach, atony of the stomach, dilatation of the stomach, etc.; all of which are either anatomical changes in the structure or form of the stomach; or, as in locomotor ataxia, it may be a symptom of the disease of some distant organ; so that, when we use the word *dyspepsia*, we use that term as it is generally understood, meaning *that food is not properly acted upon by the stomach.*

§I. PERSONS AFFECTED.

Dyspepsia is a disease mostly met with in the easier conditions of life, such as with business men, editors, writers, journalists, artists, reporters, professional men, society people, and others, but anybody can be affected from a King (Louis XIV) to the humblest and poorest man ; working men and laborers are more free from it than others. It will not come suddenly except in cases of poisoning, either by strong acids or other poison. . The stomach will at first discreetly notify you by its uneasiness that it is tired, that it needs more care or better food ; if you answer to its call and are careful, the stomach will take up its work and digest very well until you abuse it again, then it will not tolerate much, and will drift into a dyspeptic state which will cause you to lead, for years, perhaps, a miserable life.

§II. NERVOUS DYSPEPSIA.

There are as many kinds of dyspepsia as there are causes of this trouble, but there is one kind that is too often overlooked, and that is "Nervous Dyspepsia." Some writers say the only cause of dyspepsia is lack of nervous strength ; others say the nervous system has nothing to do with dyspepsia.

I think both are wrong. Certainly I do not pretend to say that every case of dyspepsia has originally its cause in the nervous system, but "every case of dyspepsia has more or less nervous weakness attending it, and that has been too long overlooked; tone the nervous system, put it in a state where it can better resist disease, and nervous dyspepsia or any other kind of dyspepsia will disappear, or improve." Ewald says in his book on diseases of the stomach: "Of course there are exceptions to the well known rule that people attacked with the gastric neuroses are usually those who live in large cities, and especially those better situated, whose struggle for existence demands an especial expenditure of labor and mental excitement to keep up with the demands of an advanced culture. I have seen quite severe cases of "nervous dyspepsia" also in farmers, working people, factory girls, and finally where one would least expect it in sailors."

We have seen in Part I, Chap. I, Sec. III, that the terminal branches of the right and left pneumogastrics, and the sympathetic all go directly from the brain to the stomach.

We can therefore well understand why the least trouble in the stomach will be *telegraphed* to the brain which will be affected (nervous headache). Also if the brain is too active and calls forth for its own use too much nervous force, we know that the first organ to be weakened and troubled is the stomach.

Nervous dyspepsia does not announce itself in a brutal or loud way; it is almost fully established before one is aware of its presence, and that is what makes nervous dyspepsia so dangerous, for it is noticed only after a considerable drainage upon the nervous system has been made, weakening thereby in a serious way the whole individual before he is conscious of it.

It is easily conceived that if the brain, central nervous organ, works more than it ought to normally, it must drain other organs of their nervous force to provide for its own needs, thereby monopolizing a great part of what these organs require for their own functions. The stomach communicates directly with the brain, through the nerves, and the consequence of a long continued effort of the brain is a correspondent debilitation of its nervous action upon the stomach, so that digestion is less thorough, takes a longer time—and the way is clear for all kinds of dyspeptic troubles.

Dieulafoy says: "When we think that the two agents of digestion, (movements and secretions), cannot be produced without the co-operation of muscles, of glands, of blood vessels, of nerves of motion and of sensation; and that it is sufficient that only one of these composing parts should be impaired in its functions, for the whole digestive action of the stomach to bear its effects, it is easy to see the numerous causes by which dyspepsia can be brought about.

But it is well to remember that the muscles, glands, blood vessels, etc., are *all under the power of the nerves*, and then one will realize how important the action of the nervous system is over digestion.

Hartshorne in his "Essentials of the Principles and Practice of Medicine," at the chapter on Dyspepsia, says: "Insufficient or perverted innervation may originate or intensify any or all of the morbid states or actions of the stomach. Sometimes this is so obviously primary and predominant as to justify the term of Nervous Dyspepsia."

§III. CAUSES OF DYSPEPSIA.

As I have said before, there are many causes of dyspepsia, but you will notice that in most of the cases there is a thread connecting the nervous system with dyspepsia.

Any long continued brain work which requires a decided concentration of the intellect is a source of general weakening, and the stomach, which is constantly working to feed the other organs, will grow weak quicker than the others. In every instance of general debility, caused by disease or by anything else, the stomach must be carefully looked after. The mental strain which business men undergo is no uncommon cause of dyspepsia; stage people are very apt to become dyspeptics; men of letters, editors, painters, writers, professors of music, professors of languages, lawyers, etc.; in short, all those whose profession keeps them indoors *are candidates to this disease.*

Imperfect mastication caused by a bad set of teeth, or also when a few teeth have been extracted and not replaced by artificial ones, is no uncommon cause of dyspepsia.

We know that if a grindstone is not in perfect order, the flour will be coarse, the same with our food, if it is improperly ground by our grindstones (teeth), the stomach will have to work much harder and finally will refuse to digest such imperfectly masticated food. Sleeplessness will bring on dyspeptic spells, great surprises, good or bad, financial losses, the loss of beloved ones, etc. are, often unthought of causes of dyspepsia.

Too little or too much exercise, too much fatigue, excessive study or mental excitement, inordinate use of ardent spirits, opium, tobacco, coffee, etc., are as many causes of dyspepsia. There is also a special kind of dyspeptic trouble with hysterical persons. In such cases antispasmodics will be of great service besides the ordinary dyspeptic treatment.

Very often dyspepsia is a nervous manifestation of "overwork" with society people. The strain on their nerves is very great; they receive, make visits, go shopping; in the evening they eat after returning from the theatre; when "en soiree" they take ice creams, frozen puddings, "des glaces," etc. In short, the high pressure conditions of life are so exacting that dyspepsia is sure, soon or late, to befall those whose brain is over-stimulated, whose stomach is abused, and nervous forces are over-strained, so that exhaustion and decadance are inevitable.

Dyspepsia is often a *reflex symptom* of the unhealthy state of another organ, such as the *brain*, *spinal cord*, but more often of the *liver*, *kidneys*, *ovaries*, *tubes*, *womb*, etc.

Chronic disorders of the female as well as of the male sexual organs may be followed by chronic dyspeptic conditions. I would here say, that the normal process of menstruation causes relaxation of gastric digestion, or even complete absence of free hydrochloric acid in the stomach contents, as was first demonstrated by Kretschy, and later confirmed by Fleischer, Boas and Ewald. How much greater reflexes will be referred to the stomach and intestines by amenorrhœa and dysmenorrhea, the climateric period and chronic disorders of the uterus which are associated with an irritability, or even with a direct excitation of its nerves! Hence, we can understand why H. Kirsch found "dyspepsia uterina" most frequently in retroflexion of the enlarged uterus than in malposition, in general myomata, pelvic exudations, with traction on the uterus and its adnexa, ulcers of the cervix, and ovarian tumors ; such dyspeptic conditions which may have persisted for years have been cured in a surprisingly short time by appropriate *local* treatment.

Very often a liver trouble is marked by no other sympton than a digestive disturbances, this is especially true with bilious persons.

Lastly, medicine, out of place, has many times been the unthought of cause of dyspepsia.

CHAPTER XVII.

DIAGNOSIS AND PROGNOSIS.

§I.

DIAGNOSIS.—The diagnosis of dyspepsia can be made by any one, but what must of all necessity be found in the cause of it, so that once removed, dyspepsia will disappear; whether it is caused by a cancer in the stomach or an ulceration either tuberculous (scrofulous) or simple, by a disease of the spinal cord as locomotor ataxia, by heart disease, by gout, by Bright's disease, by tubercolosis (consumption), or whether dyspepsia has a nervous weakness for origin, as in eight out of every ten cases, and lastly whether caused by improper nutrition, etc., dyspepsia is very easily distinguished from other diseases, but it belongs to a specialist to discover the cause and to cure that disease so exasperating known as dyspepsia.

§II.

PROGNOSIS.—Dyspepsia is essentially a chronic disease (acute dyspepsia is called indigestion and lasts but a few days) ; the longer one has had it, the farther away is recovery. Recovery will be gradual with many *rechutes* before a cure can be obtained, and even then the dyspeptic must take great care what to eat and how the food is cooked ; he must eat very slowly letting the teeth do their function, *i. e:* make a *paste* of whatever food is put into the mouth, so that the stomach will not have to do the work of the teeth.

To masticate thoroughly is an absolute rule for all dyspeptics.

The work of reconstruction must begin prudently and slowly, the greater the weakness of the nervous system, the more closely must the gradual treatment be followed, as the weakness is a consequence or a cause of the decadent state of the whole individual.

Persons affected with a heart disease must always be very cautious, for it is not uncommon that such persons will die after a hearty meal.

With a severe regime, dyspeptics will outlive stronger persons than they, and will die only after involuntary imprudence.

CHAPTER XVIII.

SYMPTOMS.

Many times two or more of the following symptoms are present in the same case of dyspepsia—seldom will there be only one symptom. Each individual will suffer in his own way as he is unlike his neighbor and does not interpret the symptoms in the same way, some are more sensitive than others, and will consequently suffer more with relatively less dyspepsia.

The first symptoms usually pass unnoticed until the attention is drawn to them; the patient does not sleep well, and dreams often; he is more nervous than usual, more irritable, will cry at the slightest cause, and is restless. He will have attacks of dizziness, and he *feels* his stomach.

When food arrives into the stomach, there is a heavy feeling in the pit of the stomach; during digestion, which takes a longer time than usual, the mouth is clammy and has a queer, sour taste; there is a general uneasiness, the head feels heavy, the eye balls are sore and sometimes there is an almost invincible tendency to yawn or to sleep. The face sometimes is flushed up, or sudden flushes of heat will be felt; there are pandiculations; the thirst is generally increased, and we find that the appetite is always modified, either increased, diminished or perverted.

Many will feel a general sensation of lassitude, others will be troubled with severe dyspnœa, which sometimes will go so far as creating a sensation of suffocation; other times very severe cases of neuralgia will, after all kinds of treatment, disappear when the dyspeptic trouble has been attended to.

A pain or an ache, more or less sharp, in the pit of the stomach, is most always felt after eating; some feel bloated and will belch up gas. The gas is either odorless or has an offensive, sour smell, with a disagreeably rancid taste; the amount of gas belched up is sometimes fabulous. Others will vomit when the food isnot properly digested; this last symptom is especially frequent in children or infants, the milk is vomited all curdled and gradually green diarrhœa will set in; finally the poor little ones will be but the living pictures of a skeleton.

When mesycism* is present it is due either to a neurosis or to a malformation of the stomach.

---

*Disease or affection in which food after a stay more or less prolonged in the stomach is brought back up into the mouth to undergo another new elaboration and once more to be swallowed after the way of the ruminants. AUTHOR.

Observation IX. About two and a half years ago, I was called to see a Mrs. M—— who was said to be dying of heart disease; numerous physicians had told her that such was her trouble; after a very careful examination and listening to the history of the case I pronounced it "Nervous Dyspepsia" to the surprise of the patient and of a whole family. They were all a little incredulous, the patient however (liking my diagnosis the best she had heard of heretofore) was willing to try my treatment. For sixteen weeks she followed a regime, also took two treatments a week of static electric sparks, and has been well ever since all dyspepsia and pain disappearep.

Another symptom of dyspepsia is pain in the heart, this however occurs infrequently. I will here give the relation of a clinical lecture about a case of digestive reflex neurosis of the heart by G. A. Fackler, of Cincinnati, clinical lecturer on diseases of the chest, at the Cincinnati College of Medicine and Surgery.

"The case which I am about to present to you is one of the class of cardiac affections, which, because of the fact that the manifestation of symptoms is transitory and especially apt to occur in individuals in robust health, is regarded by the majority of general practitioners as of purely functional character. He is justified in this assumption because of the negative results of a physical examination. He, as a rule, is in a position to assure the patient that he can discover no valvular lesion or abnormality in the size of the heart, and is inclined to regard the applicant as a hypochondriac.

The patient is a man, 30 years of age, with an exceptionally good family history. His occupation as an accountant necessarily places him among those who are compelled to lead a sedentary life. The first, most prominent and recurrent symptom observed by the patient was apparently connected with the respiratory or circulatory organs. Six months ago, without any premonitory signs, he experienced a sensation of oppression, accompanied by fear. Especially noticeable was the severe, inexplicable dyspnœa, which was accompanied by cardiac palpitation. These attacks have been of frequent recurrence, and gradually, in the history of the case, the cardiac symptom assumed prominence and controlled the general picture. We are unable to secure any evidence leading us to suspect gastric disease. With the exception of slight distention and sensitiveness in the epigastric region, there are no subjective or objective signs of abnormal gastric digestion.

In order to avoid unnecessary repetition, let us consider the class of cases of which this is a striking example, and to which the term digestive 'reflex neurosis of the heart' has been applied. The clinician is justified in assigning these cases to a special category, because of the series of characteristic symptoms which accompany the course of the affection.

The majority of patients are men, between thirty and forty years of age, and whose occupations or pursuits preclude bodily exercise. The initiatory symptoms are referred to the respiratory or circulatory functions. As in the case before you, the onset is sudden and may be described as an asthmatic attack. From its very inception the attack is complicated by tumultuous cardiac action. Frequently the patient, who is anxiously noting the pulsations, observes striking variations in their force and occasional intermittent action. Hence, in his description of the ailment, he dwelt upon the heart disturbance to the exclusion of other details. In the first stages of the disease, a careful inquiry as to the possible existence of any gastric disorder furnishes negative results.

The duration of the attacks is variable. Before the affection assumes a chronic character, the attacks last two or three days, and are as sudden in their disappearance as they were in their onset. In contrast to the apparent severity of the illness, the patients are, as a rule, well nourished, healthy-looking individuals.

The intervals between the attacks growing shorter, the chronic stage is gradually entered upon. Now we are dealing with a very obstinate, complicated process, the most disagreeable feature of which is atonic dyspepsia, involving the entire digestive tract.

As may be expected, psychic depression and hypochondriacal symptoms are features of the course of the disease.

All observers agree that the prime etiological factor in the production of this state is a dietetic error. In all probability we are dealing here with a peculiar idiosyncrasy of the stomach. Vegetable acids, certain fruits, nuts, alcohol, especially in the form of acid wines, have been found to act as iritants and causes of the attacks just described. Especially, however, has it been demonstrated that the ingestion of large quantities of cold fluids, as ice water, cold beer, etc., may be regarded as an etiological factor in many cases.

When we view the co-existence of the symptoms, viz., dyspnœa, palpitation, pronounced arhythmatic cardiac action, disturbed digestion, we must conclude that the condition is one of reflex irritation of the vagus,* depending upon some injurious effects produced upon its gastric branches.

In arriving at our diagnosis we must differentiate between this and gastralgia, gastric catarrh, nervous dyspepsia. We can readily arrive at a definite conclusion if we bear in mind the severe lancinat-

*The pneumo gastric nerve is called Vagus by many writers.,

ing pain of gastralgia, the nausea, and vomiting of gastric catarrh, and the headache, somnolency or other nervous symptoms, which immediately follow the ingestion of food in nervous dyspepsia. In the cases of reflex neurosis, which I desire to distinguish from other similar affections, the diagnosis rests chiefly upon a functional disturbance of the circulatory apparatus due to irritation of a healthy normal stomach.

The prognosis as to cure is excellent; and as to time necessary to affect such a cure, it depends upon the conditions found in the individual case, and the treatment followed."

Another symptom of dyspepsia which I have seen mentioned by no medical writer is this one, better illustrated by the history of the following case:

Observation XXI. A young man of 27 years of age, an M. D., came to me in 1892, having been troubled for the last two years with a heaviness in the head, recurring more and more often, lately three or four times a week; there was no heartburn (pyrosis), no gas was belched up, the bowels were not quite regular, there was no vomiting, in fact hardly a symptom of sickness except this one. Almost every night, he felt a soreness all over the abdomen; pressure or a change of position would neither increase or diminish the soreness, it would be felt from one or two o'clock in the morning till half an hour after arising; this soreness caused uneasiness and sleeplessness. The meals were taken regularly and did not distress or cause any trouble, apparently digestion was good. In short there was not a symptom we ordinarily find pointing towards dyspepsia.

This M. D. had a large practice and was always rather busy, therefore calling too much upon his nervous forces, the result being "nervous dyspepsia in the bowels."

And what proved this most conclusively was that in 14 weeks without medicine of any kind, he was cured by no other treatment than the supplying of his nervous system with positive electricity through the static machine; and to-day, 28 months since, he is as well as can be and has not been troubled with that most uncomfortable soreness.

With some persons dyspepsia will be characterized by belching and acid regnigitation, sometimes so acid that for hours a burning sensation is felt in the throat; men who smoke too much or who drink are especially affected that way, and also those of a bilious temperament.

There is a form of nervous dyspepsia, those very nervous or of hysterical tendencies have, when craving hunger takes the place of want of appetite. The patient has the continued sensation as of a *vacuum* in the stomach, hunger is always present and when it is satisfied, it is accompanied by a sensation of faintness. This dyspepsia is generally not associated with belching, flatulency or constipation, diarrhœa is most always present.

There are certain special cases of paroxysmal dyspepsia known as "nervous gastroxia,"* and which might well be nothing but a variety of migraine; but the description of which well finds the place here. The paroxysm suddenly comes on most always after continued and excessive intellectual work, or after often repeated nightly toil, they usually come on every other month or every month; between the paroxysms the health is good.

The paroxysm is constituted by a violent headache, by a burning pressure in the stomach and by vomiting so acid that for hours the pharynx will keep that hot and biting sensation; these paroxsyms last from one to three days. A few glasses of hot water, absolute quiet in bed, at the moment of the attack, is the best treatment; between these attacks, static electricity applied over the stomach and head will cure if continued long enough.

Sometimes dyspepsia will cause disorders of the vision, as diplopia (seeing double) strabismus, which will soon disappear under proper treatment of dyspeptic trouble.

A foul breath is occasionally the only symptom

---

*Described by Rossbach under the name of "Gastroxynsis." Lepine gave it the name of "Nervous Gastroxia" in the Bulletin de la Societe Medicale des Hopitaux, 10 Avril, 1885.

of dyspepsia; but one must not be misled by the foul breath due to a decayed tooth, to ozena, to catarrh, etc.; the food not being properly digested in the stomach a commencement of putrefaction sets in which causes the breath to become foul.

As everybody will notice in dyspepsia, the skin is more or less sallow, and always rough or "greasy" as I was graphically told one day by a lady.

Many cases of eruptions* such as exanthema papula, vesicula pustula, eczema, etc., have been cured by the treatment of the dyspeptic trouble which caused these eruptions.

A slight, dry, hacking, tiresome cough is often present in dyspepsia, and its only symptom; this cough will disappear as soon as dyspepsia is cured, or attended to.

The pulse is small and weak, thread-like, sometimes intermittent, and this irregularity of the heart's action is felt by the patient as palpitation. Some patients have a certain characteristic odor which is also communicated to their underwear, and with each exacerbation, this odor becomes stronger. Constipation is most always present and is sometimes the result of dyspepsia, other times, the cause of it. The fæces will usually come out in hard and dry little lumps; which is a sure sign that INTESTINAL DYSPEPSIA is present.

---

*L. Brocq, Traitement de Maladies de la Peau, Paris.

All the above symptoms will not be found in all cases of dyspepsia, nor even in the majority of them, sometimes one, sometimes another symptom will predominate and characterize the dyspeptic trouble.

CHAPTER XIX.

TREATMENT OF DYSPEPSIA.

The treatment of dyspepsia is all important, and we will divide it into hygienic, medical and electrical treatments.

HYGIENIC. Temperance in the food and drink has always been considered the best and the most natural means of enjoying good health.

Regular habits and sobriety constitute a guarantee of long life; starting from this principle, the number of meals, the amount of food and drink must be regulated by age, by the appetite, and by the digestive power of the stomach. The one that grows eats oftener than a middle-aged person, the active farmer more than the city man, etc.

Here are a few rules which, if closely followed, will bring back strength to the stomach, if one is dyspeptic, and keep it strong if one is well:

I. The meals are to be taken with great regularity, and at great enough intervals that the stomach may have time to thoroughly digest every morsel of food. It is better to eat moderately three times a day than to eat abundantly twice.

II. The meals are to be taken slowly, so as not TO PACK the stomach. RAPID EATING IS SLOW SUICIDE.

III. Mastication must be thorough, our teeth have their usefulness, they were not placed in our mouth only as an ornament, but also to perform their function, which is part of digestion; food is cut up by the teeth, and the saliva penetrates every particle of it, thereby aiding in the work to be done by the gastric juice in the stomach. Here is the invariable rule I give: "Put into the mouth a bit of solid food, then bite it 33 times, and by the time you get to the last stroke the food will be so well cut up that there will hardly be anything left in the mouth to swallow. This takes but very few seconds."

IV. Not a single morsel of food is to be allowed to pass whole or semi-solid.

V. Drink very little during the meal. If mastication is thorough there will be little need of drinking. ONE USUALLY DRINKS NOT ON ACCOUNT OF THIRST, BUT TO WASH THE FOOD DOWN.

VI. Avoid ice cream, ice drinks, etc., at meals or when out; a bit of pure water (spring water preferred) with claret, port, sherry or some light wine, about a wine glass in a tumbler of water is the best drink to take. One such glass is plenty of liquid for any one meal.

VII. Strong tea or coffee is responsible for a

great many cases of dyspepsia. Tea or coffee will do for one stomach, but will injure a dozen others perhaps, even when taken weak. Very strong tea or coffee is not to be taken at all. Of the two, coffee is the most stimulating and the most injurious if taken in large quantities.

VIII. It is almost needless to say that *appetizers*, appertizing drinks, etc., are extremely dangerous, on account of the artificial hunger created, the stomach will be over-loaded and will grow weaker by overwork until dyspepsia finally sets in.

IX. The choice and the variety of food are two very important factors in the treatment of dyspepsia. It is not wise to eat always of the same food.

X. For a hard working man, meats are required; steaks, roasts of pork, of beef, of veal, of mutton, etc., in short, eatables containing a deal of substance; others ought to eat only such food as they can easily digest, as chicken, beef, turkey, turbot, flounder, sole, rice, lentils, fruits with juicy pulp, for those that are farinaceous are of no easy digestion.

XI. It is very imprudent to eat immediately after great exertions either physical or mental, such as violent manual work or training, continued writing, speaking or singing, after hearing very good or very bad news, etc. It is better to wait until the nervous system has calmed down somewhat. It is well after such exertions, to sleep, for sleep is the rest of a tired nervous system and the time of its recuperation.

XII. For a half hour to an hour after eating, all physical or mental work ought to be left alone. A walk, agreeable conversation, laughable stories, a cigar (not for women!) after the meal are of great help to digestion. It is accelerated by cheerfulness, but this does not occur till the close of the meal, nor until the fluids are absorbed, or solidified as in the case of milk.

XIII. It is better to eat in company than alone. The solitary eater is more apt to become dyspeptic than the social one or those who eat and hold conversation between the courses.

XIV. Let pastry severely alone, it is only too true that New England eats too much pastry and the sarcastic remark of a writer about the "Pie Belt" of the United States is quite true; more dyspepsia has been caused by pastry than by all the other kinds of food put together. All kinds of pastry will lay heavily on the stomach, the only pastry not injurious is that of the variety called "sponge cakes."

Preserves, cakes and pies must be avoided.

XV. Dyspeptics must not forget this general rule, never to fully satisfy their appetite, but to stop as soon as they feel the first sensation of satiation, and to allow sufficiently long intervals to intervene between meals.

XVI. Exercise daily in the open air is very important to a dyspeptic. The bicycle is a useful, pleasant, healthful and very practical way to get regular exercise which helps to make an active brain and able body. So is bathing, which maintains the healthy action of the skin with which the stomach sympathizes. But active exercise ought not to be taken just before or just after a meal. "After dinner sit awhile."

XVII. The dyspeptic may be able to do the most for his own cure. In the words of the late Professor N. Chapman: "If he be intemperate, he is to be sober; if he uses opium or tobacco, he must relinquish it; if indolent, he must be awakened to enterprise; if luxurious, he must consent to change his scheme of life; if studious to abandon the midnight lamp; if afflicted we must cheer him with the light of hope, or, if this be difficult, give him the great consolation of occupation, interest, employment."

MEDICAL TREATMENT. FIRST. The best known dyspepsia medicine, also the most used in and out of place is PEPSIN. The administration of pepsin ought to be restricted to those cases in which its *absence* from the gastric juice can be proved, that is, to cases of advanced mucous catarrh and atrophy. Here is the best way of taking it: Pepsin V to XV grains dissolved in water acidulated with hydrochloric acid, 15 minutes after eating. Fairchild's *glycerinum pepticum* in usual doses is about the best preparation of pepsin, for it contains no sugar of milk as most of the artificial preparations of pepsin do. Dujardin-Beaumetz's *Elixir peptogene*, consisting 10 parts of dextrin, 20 of rum, and 180 of sweetened water is also a very good preparation.

SECOND. Hydochloric acid is of the greatest importance in the treatment of chronic dyspepsia because it not only replaces the deficiency in the secretion and forms acid albuminates so essential for peptonization, but also because it prevents organic fermentation, or lessens it when already present. Javorski has shown that considerable quantities of hydrochloric acid may be introduced into the stomach without harm. A watery solution, as sour as the patient's mouth will tolerate, can be taken three times a day, about fifteen minutes after meals. A glass tube should be employed as the prolonged use of the acid effects the teeth.

THIRD. The lavage of the stomach has acheived the greatest success in chronic dyspepsia and in dilatation of the stomach. As this treatment must be given by a physician, it will not be gone into any further.

FOURTH. Bitters and carminatives are very important in the treatment of dyspepsia. Here are a few prescriptions for dyspepsia :

Number 1. Tinct. nuc vomicae, 2½ drachms.
Decoct. condurango, 10 ounces.

M. Sig: A tablespoonful three to four times daily, half an hour before taking food.

Number 2. Tinct. Belladonnæ, 3½ scruples.
Tinct. Nuc. Vomicae, 2½ drachms.
Tinct. Castor. Canadensis, 2½ drachms.

M. Sig: 15 to 20 drops five times daily.

Number 3. Tinct. Nuc. Vomicae, 4 drachms.
Tinct. Gent. Co., 2 ounces.
Tinct. Columbae, 1 ounce.
Aqua Menthae pip., 4 1-2 ounces.

M. Sig: One teaspoonful half hour before meals.

The three preceeding prescriptions are used in nervous debility and aggravated dyspepsia.

Number 4. Tinct. Gent., 1 ounce.
Tinct. Quassiae, 1 ounce.
Tinct. Columbae, 1 ounce.
Decoc. Condurango, 5 ounces.

M. Sig: One dessert-spoonful half an hour before meals.

Number 5. Fl. Ext. Hydrastis Canadensis, 2 ounces.
(Merrill's.)
Glycerine, 2 ounces.
Aqua Distillata, 2 ounces.

M. Sig: One tea-spoonful in a wine glass of water, half an hour before meals, and at bed time.

The two preceeding presciptions are used as a tonic and for dyspepsia in weak persons.

Number 6. Pill of Carbonate of Iron, (Vallet's Mass.), 2 scruples.
Sulphate of Quinine, 1 scruple.
Alcoholic Ext. of Nux Vomica, 5 grains.
Mix and divide into 20 pills. Take one twice daily.

In prolonged chronic dyspepsia, general debility, this will give surprising results

Number 7. Ext. of Gent., 1 drachm.
Powder of Rhubarb root, 1 drachm.
Mix and divide into 40 pills. Take 1 or 2 thrice daily.

In dyspepsia, flatulence or tendency to colic.

Number 8. Ext. of Gent., 30 grains.
Powdered Rhubard, 30 grains.
Blue Mass., 4 grains.
Oil of Cloves, 4 drops.
Mix and divide into 20 pills. Take one three or four times daily for a few days.

To prevent recurring bilious headache.

Number 9. Subnitrate of Bismuth, 1½ drachms.
Pepsin, 1½ drachms.
Sulphate of Strychnia, 1 grain.
Compound Tincture of Cardamon, 4 ounces.
Mix. Take one teaspoonful thrice daily in a wine glass of water.

In nervous dyspepsia.

Number 10. Dilute Nitro-Muriatic Acid, 30 drops.
Tinct. of Nuc. Vomicæ, 1 drachm.
Comp. Tinct. of Gentian, 4 ounces.
Take one teaspoonful in water after each meal.

In nervous dyspepsia.

In England they are fond of this bitter preparation.

Number 11. Ipecac, 10 grains.
Ext. Nuc. Vomicae, 10 grains.
Mix and divide into 30 pills. On pill half an hour before meals.

CARLSBAD and MARIENBAD waters have a world wide reputation, and yet patients, with pronounced neurosis of the stomach, ought never be sent to drink these waters, for they will increase the irritative manifestations of the nervous depression. For these nervous patients we must recommend a general tonic treatment, any of the above prescriptions, which will vary with the individual will answer.

MILK is often very refreshing. Some persons do not digest milk with ease, the same with cocoa. A cup of hot milk, before going to bed, or at one's meals, is most always surprising in its good results upon the tired stomach and head.

CARDIALGIA, a pain limited, as its name indicates, to about the situation of the cardia. It seems to depend mainly upon acidity, aggravated perhaps by butyric fermentation. Aromatic spirits of ammonia, tincture of ginger, camphor water, chloroform in five or ten drop doses, or antacids after meals, as bicarbonate of sodium five to ten grains, bicarbonate of potassium two or five grains, or a dessert spoonful of lime water will contribute much to the comfort of the patient. Carbonate of magnesia, charcoal, etc., are also taken with good results.

GASTRALGIA or GASTRODYNIA, or pain all over the stomach. "Stomach-ache" is common in dyspepsia. Carminatives are appropriate for it. One of the best is oil of cajuput, five drops at a dose, on a lump of sugar. Spirits of camphor, compound spirits of lavender, compound tincture of cardamon, and essence of ginger (Jamaica ginger) are among the most popular preparations for its relief. A mouthful of very hot water will sometimes quelch the pain.

PYROSIS, or heart-burn, is best treated by mild astringents, as oil of amber, catechu, krameria.

The following is considered by the author as one of the best prescriptions for heart burn.

Ammonio ferric alum, 3 scruples.
Cinnamon water, 6 ounces.
Dissolve. Take a dessert spoonful four times a day.

Dr. Lawson considers *the sulphites* to be almost infallible in the treatment of pyrosis; tincture chloride of iron has also been given with very good results. Sub. nit. of bism. 15 grains 3 times a day is also highly recommended.

In the *nervous diseases of the stomach*, the question will depend whether they are of an irritative or depressive nature. For local hyperesthesia, gastralgia, vomiting, and even spasmodic conditions in very nervous or hysterical patients, Ewald gives the following:

Morphinæ Hydrochloratis, 3 grains.
Cocainae Hydrochloratis, 4 grains.
Tinctura Belladonnae, 2 drachms.
Aqua Amygdalae Amarae, 6 drachms.
M. Sig: Ten to fifteen drops every hour until relieved.

When patients object to cocaine or morphine, a three to five per cent solution of chloral may be taken in teaspoonful doses every hour or two, it has a good sedative action. Massage of the stomach has also given good results.

The internal stomach douche has (in the clinic of Prof. Kussmaul) given suprising results.

Rosenthal, Leube, Rosenbach, Vizioli, Dieulafoy, Germain See, etc., have all derived splendid results from the constant (electrical) current. The induced current has also been used to great advantage. The negative pole on the lumbar region, the positive pole over the stomach. Good results have been obtained in Paris with the negative pole *in* the stomach and the positive over the epigastrium.

When nervous dyspepsia is due to a central cause, either of the salts of potassium, sodium or ammonium may be used, affording great relief; from 20 to 30 grains three times a day. The tincture of the chloride of iron is used very extensively by the author and with very good results, beginning with 10 or 15 drops thrice daily, gradually increasing to 30 drops in a wine glass of water with the white of an egg beaten in. Not one in a hundred sensative stomachs but will bear this albuminate of iron.

CONSTIPATION. Constipation is the cause of great discomfort. The effects of non-evacuation may be intestinal obstruction or inflammation, sympathetic headache, stomach or liver disorder, urino-genital irritation, offensive perspiration, contamination of the blood, etc. Consequently constipation must be cured before we can hope to cure dyspepsia when this trouble is present.

Stewed prunes, I consider a specific to *clear* the bowels. This is the way to take them: take four stewed prunes or more before going to bed, also take them once or twice in the day at meal time; there is also one rule to follow, which is, to have a *regular hour every day to empty the bowels.*

Of medicines, rhubard pills, podophyllin pills, colocynth, aloes, etc., are given when there is especial torpor. Glycerine suppositories once in a while are useful. However, the following rule has always given me good results.

The rectum is emptied, or *teased* to empty itself, every day at the same hour, and in from four to six weeks the bowels will move every day. The first attempts are not always successful, but if constant, one will be rewarded by the regular action of the bowels after a time.

ELECTRICAL TREATMENT.† These two preceeding treatments are the principal ones used in dyspepsia by medical men, *but there is one other treatment which can and ought to be used in every case of dyspepsia*, a treatment which is sure to give good results if only used properly. I refer to the special treatment with the static machine* to STATIC ELECTRICITY.

I have used static electricity in 49 different cases of dyspepsia of from a few weeks to 15 years duration, and invariably with good results.

The machine I use is a machine of French make‡

†This treatment I have been the first to use in cases of dyspepsia, and always with good results. THE AUTHOR.

*For further reference see appendix.

‡Gaiffe fils, the celebrated electrician of Paris, being the maker.

which can give me sparks from one to 14 inches in length, and in all kinds of weather.

If a weak or nervous person has dyspepsia, I connect her with the positive pole, the patient being on an isolated stool, and draw from her a spark with an appropriate instrument. These sparks have a very great tonic effect upon the nerves and muscular tissue, it is especially tonic with asthenic patients and in chronic dyspepsia. It regulates the circulation, rendering it more active, and distributing more equally the blood supply all over the body.

THERE IS ABSOLUTELY NO DANGER WITH THIS KIND OF ELECTRICAL TREATMENT, BUT IT IS ONLY A SPECIALIST THAT CAN USE IT DISCRIMINATELY.* In mild cases of dyspepsia I will use the pointed electrode, bringing it within an inch to an inch and a half of the patient, that is, until I draw a spark from the patient's body. These sparks are drawn from all parts of the body, so that all the muscles will be made to contract and thereby activate circulation. Immediately after each treatment the pulse is stronger and more regular than before.

From over the stomach are drawn most of the sparks, increasing their force as the benefit is

---

*Prof. Fleming of the Royal Institute, London, England, has recently, this month (January, 1895) proved by experiments that a static spark, although of great tension, has little in quantity. THE AUTHOR.

greater, and as the dyspeptic trouble grows better; sparks are drawn from over the liver when it is sluggish, so that its action will be more regular. The intestines are more active after a treatment over them.

The spinal column ought to be attended to at every seance, as it is a *nerve centre*, and it must be made stronger every day.

The circulation being made more active, the blood supply is greater and the nerves grow stronger at every treatment, as they are fed more, so that the beneficial effects of static electricity are rapid. Every week the dyspeptic notices that his strength is coming back, that food does not distress as much, that digestion is easier, that more food can be taken, that his weight is increasing, etc., *and this without any medicine whatever.*

THOUGH IN A FEW CASES MEDICINE IS REQUIRED, IN MANY INSTANCES, DYSPEPSIA IS CURED WITH NO OTHER TREATMENT THAN THE ELECTRICAL.

Here are a few of the cures obtained by this treatment.

Observation VI. Mrs. L——, 26 years of age, had enjoyed very good health up to 21 months before she came to me, the second day of July, 1892. Since then food would lay heavy in her stomach, she complained of great lassitude and of cramps in the stomach. There was a complete loss of appetite. The face was pinched and the complexion sallow. The appearance was bad and the general weakness increasing, the cramps were not constantly present at the time of taking food.

There was no localized pain over the stomach; no acidity in the stomach; she would occassionally belch up gas; the pulse was tense and rapid; the temperature was normal, no tenderness over the ovaries. She was rather nervous. Medicine had never helped her, she said.

The electrical treatment was begun on the second day of July, 1892, and it was given twice a week; at first the positive electricity was given very weak, then gradually, as Mrs. L—— grew stronger, the sparks being drawn from over the stomach and bowels, and along the spine, the dose was increased every week until she was well. It took about three months and a half to cure this patient, who is still well.

Observation XIX. May, 1893, Mr. J——, 42 years old; for over two years he has complained of acid eructations, of constantly loosing appetite, there was most always present a distention of the abdomen, with pressure and fullness there as gases were present in the bowels. He was sleepless and the head ached often, there were spells of dizziness, the bowels were irregular and costive, anxiety was pronounced, he was tending to hypochondriasis. The tongue was coated, the stomach of normal size; *but by the Ewald tests with iodide of potassium and salol* both showed retardation of absorption and motion. This patient was given four ounces of the following solution, besides the electrical treatment over the stomach, head and along the spine.

Tinct. Nuc. Vomicae, 1½ drachm.
Decoc. Condurango, 4 ounces.

M. Sig. One teaspoonful half an hour before meals, with or without water.

This man was cured in 14 weeks and since then dyspepsia has not troubled him.

Observation XVI. Is given in full on page 87.

Observation XXXIX. May, 1894, I was called to examine a lady, Mrs. J—— who had been sick for fifteen years, medicines were of no avail, and her strength was gradually decreasing. I came to examine her and found her in bed, she was rather thin and feeble, and appeared to be about 36 years old. Gas was belched up quite frequently, but there was no vomiting or acid *eructations*; the stomach was found to be somewhat enlarged; palpitation of the heart would recur now and then, but there was no heart disease to be found; the liver and kidneys were normal, the lungs were examined and were perfectly sound, the bowels were irregular, constipation being present. There was a pain in the left side of the abdomen, which was due to a local inflammation she had when her only child was born in 1882, which inflammation had become chronic.

This person had suffered such pain and distress in her stomach when food was not digested that she would eat of very few foods and sparingly of them. There was a marked nervousness, but absolutely no hysterical or hypocrondrical symptoms.

The following diet was given her. The whites of two eggs beaten in spring water to be drunk as often as four to six times in 24 hours. As soon as Mrs. G. was able to get up, she came to the office, traveling 80 miles in the cars on her way to and fro and yet there was a gradual improvement under the electrical treatment which was given twice a week, all over the body. Faradisation through a thin coil was made over the left side, the positive pole on the anterior abdominal wall, the negative in the lumbar region. Stewed prunes were ordered for the constipation. The patient is still under treatment, but this is her present state. The bowels are now regular, the stomach will digest quite a good deal. Mrs. G. has grown fleshier, the nervousness has most all disappeared, the pain in the side is gone, and the palpitations of the heart have not recurred and it is an uncommon thing to belch up gas, so that we can truthfully say that recovery is in sight.

CHAPTER XX.

GASES IN THE STOMACH AND INTESTINES OR GASTRO-INTESTINAL PNEUMATOSIS.

In the normal and natural state, the intestines contain a certain amount of gas, absolutely necessary so that the walls of the intestines will not *fall in over each other*. When that normal amount is increased the person has in the intestines a noise caused by the displacement of the gases. This noise is called barborygms.

When one eats fruits or vegetables with thick skin, which is not digested, they will cause more gas to be present in the bowels than usual, but this gas will dissappear with the cause. If these gases recur often, they will determine a flatulent affection which is very uncomfortable, often painful, and which, in the end, weakens, more or less, the digestive functions. With reason, it is believed that such food as cabbage, carrots, peas, radishes, beans, turnips, etc., also improperly cooked mealy substances, will bring on gases in the intestinal tube of certain persons, according to the temperament, to the regime, and to the social condition. In fact the farmer, the workingman, the countryman, leading an active, open-air life, can eat most any *hard-to-digest food* without trouble, whilst the business man, the city lady, nervous and delicate persons, will according to their state of health, feel that the food will not digest well and that the stomach is uneasy. As the food proceeds further in the intestines, gas will form and cause great discomfort.

Gas is formed by a sort of *fermentation*, in the bowels, of indigestible food, or, in other cases, it is due to an abnormal secretion of the mucous membrane of the stomach, and of the intestines.

These gaseous dyspepsias are real *neuroses*, very obstinate, very capricious in their course, their symptoms and their duration. To-day the dyspeptic feels sure of an early recovery, to-morrow he is morose and feels despondent. Very seldom will a person vomit in this disease. Persons troubled with "flatulent dyspepsia" are pale, their tortures leave traces upon their face, the complexion is more or less sallow, they are irritable, very nervous, and often have spells of dizziness. The amount of gas passed is sometimes incredible, everything seems to ferment and make gas in the stomach and bowels.

Belching is very frequent, the gas is either odorless or has an offensive, sour smell and disagreeably rancid taste. It is frequently accompanied by the regurgitation of fluid or remnants of food from the stomach, having a very sour and disagreeable taste. These regurgitated masses often impart a burning and scratching sensation along the œsophagus.

Soon after eating, the dyspeptics feel oppressed and bloated. There is a vague choking sensation, and if a slight pressure is made on the stomach, it will cause pain.

The dyspeptics very frequently have the feeling that the food remains abnormally long in the stomach, the result of which is a decomposition in the ingesta. This produces distention of the stomach with gas, erectations of offensive gases, and regurgitation of sour and rancid masses.

The distention of the stomach in turn paralyses its muscular fibres, and causes a feeling of tension and pain; the decomposed or insufficiently digested stomach contents irritate the intestines, and the conditions thus produced are reflected back to the stomach, and thus the vicious circle, which is present in all affections of the stomach, is completed; these decompositions usually occur at night, in the morning they may be absent or only very slight, there is most always more or less severe palpitation of the heart, causing the patient to believe that he has a heart disease.

Constipation exists as a rule. In a few cases diarrhœa and constipation alternate; hemorhoides are often present and the movements are consequently painful.

As general symptoms, we notice a dimunition of of mental activity, disinclination to bodily exertion, languor during the day, especially after meals, headache or a feeling of oppression in the head, and a more irritable disposition. The dyspeptics frequently complain of a feeling of heaviness in every limb, cold extremities, itching and formication. Sleep is deeper and longer than usual, but is not refreshing and is disturbed by hidious dreams.*

TREATMENT. For the treatment of this affection see the hygienic medical and electrical treatments on page 92.

Here are a few preparations, any of which will give immediate relief when the stomach is full of gases: Infusion of aniseed, oil of cajuput, five drops at a dose on a lump of sugar; spirits of camphor, five to 20 drops; compound tincture of cardamon, one-half to a teaspoonful; essence of

---

*One can, besides the above symptoms, have one or more of the symptoms described at page 81.

ginger, five to 30 drops; compound spirits of lavender, 30 to 60 drops; essence of peppermint, five to 25 drops; tincture of capsicum, one to 10 drops, on a lump of sugar, etc., etc., etc.

Any of these will almost instantly free the stomach of the gases accumulated in it.

CHAPTER XXI.

CORRELLATION OF THE DISEASES OF THE STOMACH TO THOSE OF OTHER ORGANS.

We have to discover in this chapter the real cause of the dyspeptic disturbance, which is only a symptomatic feature of a disease *outside* of the stomach.

The effect of the diseases of other organs upon the stomach, and their reciprocal action have been carefully studied by W. Fenwick, Louis, Andral, Germain See, Dujardin-Beaumetz, Ewald, and others, and is today a well established fact.

There is always more or less marked catarrh of the mucous membrane of the stomach in the following: in diseases of the kidney, pulmonary phthisis, chronic bronchitis, emphysema, valvular lesions of the heart, etc.

TUBERCULOSIS. The most important is tuberculosis. It is well known that the course of phthisis may be marked by dyspeptic symptoms, and as Louis, Andral and Bourdon pointed out long ago, there are many cases of tuberculosis in which the first symptom to attract the attention is dyspepsia. Hutchinson says that, in 33 per cent, dyspeptic symptoms precede the onset of tubercular manifestations. These persons are delicate and anaemic, they begin to complain of loss of appetite, oppression and fullness after eating, and irregularity of the bowels, they suffer from regurgitation and a foul taste in the mouth, they feel feeble and languid. They are treated a long while before the dry, hacking cough is taken notice of and consumption suspected.

ANÆMIA and CHLOROSIS.* Thechanges in the digestive tract in anæmia and in chlorosis are closely allied. They undoubtedly play an important part which has been very much neglected; hence, in the treatment of anæmia, efforts should first be made to improve the conditions of the digestive organs and then the composition of the blood.

HEART DISEASE. In the valvular affections of the heart, the nature of the lesion causes a veinous congestion and the symptoms of a chronic catarrh

---

*Austin Flint was the first to call attention to the relation between anæmia and atrophy of the gastric glands.

of the stomach. When a patient comes to me for a heart trouble, I always examine the symptoms of the stomach, and eight times out of ten, the patient has dyspepsia pure and simple; but heart disease, by its veinous congestion of the stomach, causes dyspepsia with all its weakening effects upon the whole system, for, if digestion is poor, the patient will not recuperate his strength as thoroughly as one with a good stomach.

DISEASES OF THE KIDNEYS. Biernacki lays stress upon the retention of the *metabolic products*, which lessens the secretion of the gastric juice by emans of nervous influences.

RENAL TUMORS or CANCER, may for a long time cause only disturbances of digestion, anorexia, vomiting and emaciation.

Colleville, in the "Progres Medical, No. 20, 1883, Paris," reported a case, where these, up to the patient's death, were the only symptoms; when there is a floating kidney, we will always find that there is pain in the stomach, and that dyspeptic disturbances are at first the only things complained of.

LIVER. The liver stands in such close relationship to the stomach, that serious functional disturbances of the one, are, without exception, reflected on the other; this close connection, and the fact that so many noxious substances introduced from without act on both visceras at once,—I will only mention alcohol—render it very difficult to say which is affected first. For example, in the very great majority of cases, cirrhosis of the liver is accompanied by chronic gastritis, yet, even if we observe that the symptoms of a doubtful hepatic cirrhosis have for a longer or shorter time preceded a chronic gastric catarrh, we are utterly unable to tell whether the two stand in a casual relation or are simply coincident. Nevertheless, we should never forget the fact that many cases of hepatic cirrhosis for a long time run their course as chronic gastritis, and that the same is true of cancer of the liver.

Often have I been called upon to cure "dyspepsia" when after a careful examination of the stomach contents, and after the Ewald tests with iodide of potassium and salol, I would be suprised to find that all the dyspeptic disturbances were due to a sluggish liver.

Observation XXXVI. Mr. Cal, aged 29, had been sick for eight months previous to his call at the office, (October 10, 1894) and had been under four different physicians' care all that time. His appetite was not good, his limbs felt heavy, his head ached often, gas was belched up now and then, the bowels were somewhat costive and there was a dull ache in the pit of the stomach towards the right side. In this case *no electricity* was used but I gave him the following pill, he has taken two a day with most surprising results. His liver acted well from the start, and he is cured of his *dyspepsia* and all the other troubles.

Podophillum, 1-4 grain.
Ext. Rhubarb, 1-4 grain.
Ext. Gentian, 1-2 grain.

DIABETES. Diabetes gives rise to error most frequently. For many years diabetics are considered to be suffering from some stomach trouble, until the urine is examined on account of the development of the specific symptoms of emaciation, pruritus, polyuria, ravenous appetite, dental caries, ocular disturbances, constant thirst, etc.

GOUT. The relations of gout to the disturbances of digestion have been especially discussed in English medical literature. According to some writers, there is a specific gouty disorder of the stomach resulting from the uric acid diathesis. Thus, not long ago Burney Yeo claimed that one of the prominent manifestations of this condition was dyspepsia in all its forms.

RHEUMATIC DIATHESIS. This rheumatic diathesis has played a prominent part in French medical literature. Ewald, of Berlin, says that he has often met cases in which the dyspeptic disturbances and pain were so severe during an attack of rheumatism that the pains in the joints were comparatively insignificant.

SKIN. Pidoux,* of Paris, has paid particular attention to this subject. The same can be said of Brocq, of Paris, and of many others; it is too well known to insist upon that eruptions of the skin will be brought on, or an existing one made worse after errors of diet. We all know the relation of eczema to digestion. See chap. XV for further reference.

MALARIAL POISONING. It may be manifested as an intermittent cardialgia, or in the form of the neuroses of the stomach, which will be characterized be a certain regularity, and which can be relieved only by quinine as long as the patient re-

*Rapport de l'herpetisme et des Dyspepsies. Union Med. No. 1, 1886. THE AUTHOR.

mains in the malarial district. One not infrequently meets with such cases of intermittent dyspepsia.

These various manifestations are quite common in New York and Boston, and should always be borne in mind in obstinate cases of dyspepsia.

In the *treatment*, Warburg's tincture will be found to be especially useful in connection with the static machine.

APPENDIX.

ELECTRICITY.

A PAPER READ BEFORE THE NASHUA MEDICAL ASSOCIATION, NASHUA, N. H.,

MARCH 15, 1895.

BY

M. T. GERIN-LAJOIE, M. D., C. M

Member of the "New Hampshire Medical Society," Member of the "Societe Francaise d' Electrotherapie," of Paris; Member of the ' Societe d' Hypnologie" of Paris; etc., etc.

ELECTRICITY.

To understand how electricity can do good to our weakened muscular and nervous systems, it is indispensable to say a few words of the physiological action of electricity, whether galvanic, faradic, or static, as it passes through our organism; its influence upon our NERVES, MUSCLES, CIRCULATION, THE NERVE CENTERS, and all our organs.

Every one knows that the starting point of all researches about the contraction of the muscles by the application of electricity, was the celebrated experience of Galvani in 1786. After that many eminent physicians made discoveries which have brought electricity to its present state. Volta, Loder, Bischoff, Haller, Magandie, Franklin, etc., were the pioneers.

Our organism is a thermo-electric cell, with this differance: A THEMO-ELECTRIC CELL RECEIVING HEAT RETURNS ELECTRICITY, WHILST OUR OR-

GANS RECEIVING ELECTRICITY RETURN MOTION, SENSATION OR TROPHIC ACTION.

This theory of transformation of energy by the living organism has not even been proved by experimentation; but it seems to be in such close accord with the other physical phenomena that I think it the only true one, and one which will in time be demonstrated by experiments.

It would be injustice to wait for the exact knowledge of the physiology of the action of electricity before we use it in therapeutics.

Has not every medicine been used *as a medicine* before we knew what was its exact physiological action?

The carefully taken clinical observations of the use of electricity will tell us its indication and its contra-indication as it did for medicines; so that in time the clinical study of electricity will lead us to the thorough knowledge and understanding of electro-therapeutics.

There are three kinds (?) of electricity. The GALVANIC, the FARADIC, and the STATIC.

GALVANIC ELECTRICITY.

Galvanic electricity, or continuous current, is generated by what is known as the galvanic battery, that is, a number of cells used together or singly. This electricity causes no pain until the current is strong, then there is a burning sensation at the point of contact between the electrode (sponge or instrument) and the skin. This can go so far as to cause actual burn, and a cicatrix as a result. Even when not at all felt by the patient this electricity has a powerful tonic effect.

Galvanic electricity has different effects according to the pole used.

At the negative pole we have the accumulation of the alkalies.

At the positive—the acids.

So that, for instance, when superfluous hair is destroyed, the negative electricity must be used through the needle* because it leaves no cicatrix or depression; but the positive would.

FARADIC ELECTRICITY.

Faradic electricity, or induced current, (also called interrupted current) is the electricity generated in a cell, but between the cell and the part electrified is a small hammer which strikes a coil from 15 to 500 or more times a minute, which produces the buzzing noise heard, each time the hammer touches the coil (or reel) the electrical force runs through the part between the two electrodes of the battery.

---

* I have for four years used the platinum needle to the exclusion of every other, for platinum needles do not corrode.
AUTHOR.

That electricity is called faradic; it causes our muscles to contract more or less, according to the strength of the current and to the rapidity of the interruptions.

THE LARGER THE COIL, THE GREATER THE QUANTITY.

THE THINNER THE COIL, THE GREATER THE TENSION.

Faradic electricity has different effects according to the size of the coil used.

Thus we must employ the reel with a large (or coarse) coil in all cases where muscular energy is in fault, i. e. in paralysis due either to sickness or to a blow—in atrophy—where the blood vessels have lost their power of contraction—in cases where there has been an effusion, and where you want to effect a reabsorption, i. e. hæmatocele, etc., IN SHORT, WHERE QUANTITY IS NEEDED.

When the sympton pain, especially the nervous pain, is present, overshadowing most everything else, by all means use the reel with the thinnest coil, the thinner the better; for here tension is what is wanted. When nervousness is a marked symptom, as in nervous vomiting, nervous head-ache, etc. This coil (or the static machine) gives the best results; a current from five to ten minutes' duration seldom fails to give immediate relief.

STATIC ELECTRICITY.

Static electricity is generated by a static machine This kind of electricity is a powerful remedy in the hands of one who understands electricity.

As Professor Fleming of London (England) has proved by experiments in January, 1895, static electricity has an enormous TENSION, with no appreciable quantity.

The oldest and the most usual way of application is that by SPARK or COMMOTION.

These sparks can be taken either sitting on an isolated stool connected with one of the poles of the machine, or simply by standing near it and bringing any part of the body close enough to draw a spark (from the machine).

The quantity is never great, as when one steps down from the isolated stool the surplus electricity, which the body does not want to appropriate, goes immediately through the ground.

Therefore, THE BODY APPROPRIATES FOR ITS WANTS JUST WHAT IT NEEDS, THE REST IS DONE AWAY WITH THROUGH THE SOIL.

The beneficial effects of one electrical treatment usually last from a few hours to two or three weeks. In some cases it is sufficient to even cure, when the trouble is recent.

Some constitutions seem to lose, or *use up* in a very few hours, the electricity taken, the system does not retain it. I have known a lady of feel the beneficial effects of the electrical bath for about four hours only, and she was obliged to come after treatment every day until cured of the neuralgia in her ovaries.

Usually the effects of each treatment last three to five days or more until one is again in perfect health.

I have found that in combining both electricities, galvanic and static, the beneficial effects upon the system were greater and more apparent than when only one kind was used.

In using both electricities at once I can say:

THE QUANTITY THAT IS WANTING IN STATIC ELECTRICITY IS SUPPLIED BY THE GALVANIC, AND WHAT TENSION IS LACKING IN GALVANIC ELECTRICITY IS PROVIDED BY THE STATIC.

So that the body is receiving what comes nearest to perfection in electricity, for what is wanting in the one is supplied by the other.

EFFECTS OF STATIC ELECTRICITY.

I will say from the start that I have used the positive electrictity of the machine most exclusively.

For electricity to have a sedative effect seems at first a parodox, for you may ask WHY IS IT, if a storm will cause such a nervousness or nervous irritation with many, that the static machine from which you draw LIGHTNINGS in miniature will NOT have similar effects? Why, simply because the dose is much smaller and the atmospheric conditions are not the same, though the electricity is identically of the same nature.

In Electrotherapy the *dose* has everything to do with the effects. This is so evident that I need say no more about it.

Patients after a static bath FEEL a reaction. This sedative effect is so constant that even after the very first treatment the tired nerves feel stronger and less *unstrung*, there is a feeling of general relief, sleep is calm and restoring; sleeplessness is often cured after two to five static baths.

But if an overdose is given or if used by inexperienced hands, static electricity will cause excitation, agitation, sleeplessness, and bring on marked nervous disturbances.

It is very hard to know exactly how much electricity one person can take as the effect varies with the temperament.

For instance, hysterical persons require from 20 to 25 and 30 minutes' treatment, whilst neurasthenic individuals are greatly benefited by baths of 5 to 10 minutes' duration.

The best rule to follow is to begin with very small doses, gradually increasing them as the benefit is greater until the maximum beneficial dose is reached. The state of the patient being your BAROMETER.

The sedative and tonic effects are very pronounced in neurasthenia — nervousness — weak nerves*—WHEN YOUTHS HAVE OUTGROWN THEIR NERVES—in convalescence after a severe or protracted sickness—in sleeplessness—in dyspepsia it is most astonishing in its curative effects†—in nervous palpitation of the heart—in sciatica—in migraine—in nervous or sick headache—in neuralgia wherever found—in nervous vomiting, *also in vomiting during pregnancy*—in hysteria or where there are hysterical tendencies—in dysmenorrhea and other uterine or ovarian pains, electricity has no superior as a releiving and curative agent—in impotency it has proven itself of the greatest value—to men after a *night out*, it will restore the nervous system to its proper equilibrium, enabling them to resume their daily work with no other effects than the remembrance of their escapade. The bath will take all tired feeling out of the nervous system and restore the suppleness and free action of the muscles, etc.

---

*In all these there is a marked difference. though usually these terms are used to designate the same trouble.
DR. GERIN-LAJOIE.

†As I have pointed out in my "Dyspepsia and how to cure it." DR. GERIN-LAJOIE.

When the circulation is defective in any part of the body (cold extremeties, etc.) electricity tones the muscles and blood vessels, bringing them back to their natural elasticity and strength, so that the blood supply is again regular and normal.

When the LIVER is LOGY* a few treatments over it will render the circulation of the *portal system* more active, and regulate the secretion of the bile.

The immediate effect of the electric spark is to cause the muscles touched to contract. Muscles, which for months have been paralyzed, will after a few weeks of electric treatment be restored to their normal state of contractibility, they also will become strong and active once more, provided, however, the muscular fibre has NOT undergone FATTY DEGENERATION.

Here is the observation of an interesting case:

Mrs. M———, age about 24. About a year ago, (February, 1894,) her first child was born. During the last months of her pregnancy, her legs and feet were very much swollen, it was impossible for Mrs. M——— to wear anything but large slippers (six or seven was the number). After the birth of the child the limbs did not return to their normal state, the muscles were slow to act, and in both feet was felt a most uncomfortable cramp all day long. This lasted from June to December, when

---

*Lazy, sluggish.

her husband, an M. D., called to ask about the advisability of using the electrical treatment. I urged him very strongly to have his wife come in and take the treatment. No other treatment had given any relief.

After two months, all cramps had disappeared, not to return, the muscles came back to their normal state of contractibility, the large veins over the legs disappeared, and to-day this lady is a picture of perfect health, and suffers no more from those dreadful cramps.

No one can deny that electrotherapy has an energetic action upon the molecular exchanges, upon the phenomena of endosmosis, upon the capillary circulation and consequently upon the nutrition of of our tissues and organs.

ELECTROTHERAPY, WHEN NUTRITION IS TROUBLED, HAS THE ESPECIAL EFFECT OF BRINGING IT BACK TO THE NORMAL, AND THAT EFFECT IS NOT MISSING EVERY TIME THERE IS NO IRREPARABLE ANATOMICAL LESION.

ELECTROTHERAPY, IN EVERY DISEASE WHERE THE NERVOUS SYSTEM IS A FACTOR, WILL HELP GREATLY IF IT DOES NOT ITSELF ALONE AFFECT A CURE.

www.ingramcontent.com/pod-product-compliance
Lightning Source LLC
Chambersburg PA
CBHW021814190326
41518CB00007B/586

* 9 7 8 3 3 3 7 8 1 1 7 4 7 *